FARM ACCOUNTING
AND BUSINESS
ANALYSIS

FARM ACCOUNTING
AND BUSINESS
ANALYSIS

SYDNEY C. JAMES
EVERETT STONEBERG

IOWA STATE UNIVERSITY PRESS / AMES, IOWA

SYDNEY C. JAMES is an agricultural economist at Iowa State University in the Department of Economics. He has taught farm records and farm management classes for 13 years, 11 of which have been at Iowa State. His research has been in farm management including record-keeping systems. He holds the M.S. degree from Utah State University and the Ph.D. from Oregon State University.

EVERETT STONEBERG is a farm management extension economist in the Department of Economics at Iowa State University. He has worked for the cooperative extension service for over 27 years and has been associated with farm records over this entire period. He has also worked closely with the farm business associations in Iowa. He holds the M.S. degree from Iowa State University.

© 1974 The Iowa State University Press
Ames, Iowa 50010. All rights reserved

Composed and printed by
The Iowa State University Press

First edition, 1974

Revised Second Printing, 1976
Third Printing, 1977

Library of Congress Cataloging in Publication Data

James, Sydney C.
    Farm accounting and business analysis.

    Includes bibliographical references.
    1. Agriculture — Accounting. I. Stoneberg, Everett, 1919-   joint author. II. Title.
S567.J327   657'.863   74-16443
ISBN 0-8138-0605-4

CONTENTS

PREFACE

THIS BOOK HAS EVOLVED from several years of teaching farm records and accounts at Iowa State University as well as from experience in extension work and with farm business associations. It is written for college freshmen and sophomores, but it could be used in vocational agriculture courses at the high school level and seems particularly well suited for technical agricultural programs offered by community colleges. No prior experience with bookkeeping or accounting is necessary, but knowledge of agricultural production would be helpful. The fundamental elements of accounting are stressed; and although the single-entry system of accounting is the major method discussed, basic elements of double-entry accounting are also explained and illustrated.

The emphasis in this text is upon use of records and accounts as a tool of management. How data are organized, recorded, and analyzed is discussed. Techniques are given for analyzing individual enterprises as well as the total farm business (using the same format for each) in such a way that both strong and weak elements are detected. The use of farm records and accounts in solving management problems is illustrated in Chapter 11. The data are used to bridge the gap between the past and the future and to illustrate economic principles, budgeting, linear programming, and analytical tools of management.

Tax management is also an important part of the text; considerations that become a part of accounting procedures are introduced throughout. Tax decisions that can be separated are placed in a chapter by themselves; however, details concerning which forms to use and how to fill them out may be found in Internal Revenue Service materials.

The organization of this book results from classroom testing and differs from other farm accounting texts. Chapters treating purchases and sales of capital, depreciation methods, and credit accounts precede the chapter on the net worth statement. In this way the background is laid for discussion of materials included in the net worth and income statements. Production records are discussed following the income statement and just preceding the farm business analysis, thus incorporating both physical and financial measures. Even though the authors consider this sequence logical, others are possible. Chapter 6, "Production and Service Record," or Chapter 5, "Receipt and Expense Accounts and Income Statement," could be presented first. Also, it would be possible to break the discussion into units.

Test questions and laboratory assignments used by the authors in teaching the course can be purchased separately as a workbook that is divided into two sections. Section A presents questions and problems on the individual chapters that can be used for class assignments, class discussions, or test questions; while Section B contains several total farm problems that can be used as assignments in a separate laboratory. They should be very helpful to the instructor as well as to the student who wishes to apply the examples to practical problems. The total farm problems require the use of all the record techniques discussed in detail in the text. The longer, more complicated problems can be worked in parts corresponding to the chapters, while the simpler ones can be used for summary purposes.

The authors recommend the text as a practical guide for keeping records that are helpful to management for business and tax purposes on modern commercial farms. It should be useful to the educator as well as the practitioner. Farm business association fieldmen, county agents, vocational agriculture teachers, bankers, farmers, and others concerned with farm records and accounts should find this book useful as a reference.

FARM ACCOUNTING AND BUSINESS ANALYSIS

IMPORTANCE OF RECORDS AND ACCOUNTS

THE FOUNDATION OF ANY SUCCESSFUL BUSINESS is a well-organized set of records and accounts. Carl Malone, a leading farm management educator in the Midwest for many years, has said, "No farmer, however good his memory, can keep all the details of his business in his mind. In fact, if he tries to do so, it is good evidence that he is not a first-class manager. Successful farmers say that one of the secrets of good management is to have a record keeping system so there is no need to remember details." Records and accounts provide a major tool for sound business management; they record the heartbeat of the business. The pulse must be recorded in all parts of the business, as in the body, for it to remain alive and functioning productively. This is just as true for today's modern commercial farm as for any large business corporation. Farming today is big business and as such requires detailed records of resource use and production as well as financial accounts of the flow of money into and out of the business. A farm of sufficient size to produce a return of $8,000 to $12,000 per year to its operator (equivalent to what he could earn in industry) will probably have an investment of over $225,000 and a volume of business of more than $50,000 per year. Table 1.1 illustrates the investments and corresponding incomes of some farms in the Midwest.

A student summarized the importance of accounting and farm business analysis when he said, "The best year my dad ever had was when he was laid up with a leg injury. He couldn't do much work on the farm and was forced to spend his time in analyzing the farm business and managing it."

Agriculture in the 1970s has grown into a businessman's industry. Hard physical effort no longer ensures success. Although machines are now available to do much of the hard work, exercise of the mind brings business success. It should not be inferred that

Forms illustrating the use of concepts and materials presented in this book may be found in Midwest Farm Account Book by Sydney James and Larry Trede, 1967, available from the Iowa State University Bookstore. Other farm account books also have forms that could be used.

TABLE 1.1. Average capital investment, volume of business, and income on farms in central Iowa, 1967-71

	Size of farm (acres)				
Item	70-179	180-259	260-359	360-499	500 and over
Capital investment:					
Feed and livestock	$ 23,822	$ 26,610	$ 36,361	$ 45,533	$ 77,740
Machinery and equipment	9,699	10,377	13,755	17,509	26,520
Land and improvements	82,050	112,182	159,455	212,305	346,512
Total Capital	$115,571	$149,169	$209,571	$275,347	$450,772
Source of income:					
Value of crops	13,529	20,292	27,632	38,067	62,399
Livestock increase over feed costs	10,229	8,011	9,493	10,463	15,610
Miscellaneous	2,126	1,798	2,128	2,622	3,769
Gross Profit	$ 25,884	$ 30,101	$ 39,253	$ 51,152	$ 81,778
Net returns:					
Net farm income	9,651	12,341	16,549	21,796	33,015
Capital charge*	5,192	7,254	10,609	13,534	21,807
Return to operator labor and management $	4,459	$ 5,087	$ 5,940	$ 8,262	$ 11,208
Months of labor	13.0	13.0	14.0	16.0	23.0

Source: Farm Business Summaries for Central Iowa, 1967-71, Iowa Coop. Ext. Serv., Ames.
*Net worth capital was charged at 5% on fixed capital and 7% on all other. Interest on liability capital was included as a business expense.

records and accounts can solve all a farmer's management problems. They are only one tool to be used in connection with others such as economic principles and budgeting. The functions of management (decision making) have been defined to include:*

1. *Recognition of a problem.* Problem perception may be the most difficult task facing the manager. A problem is said to exist if there is a felt difficulty, but the real problems may not be where the difficulty is felt. For example, lack of income may be the felt difficulty, but the problems may lie deep within the business organization or even with the spending patterns of the farmer or his family. Some of the most difficult problems to distinguish are those that go unrecognized because there is no felt difficulty. The farmer's knowledge level is not sufficient to tell him that a problem exists. For example, suppose a farmer analyzes the costs and returns from his swine, finds a typical margin of profit, and concludes his performance is satisfactory. However, if he had analyzed his swine enterprise in detail, he would have discovered that his feed conversion was excellent but the price he was paying for feed was too high. His efficiency as a producer was higher than as a feed purchaser. The detection of difficult management problems poses a challenge to all farm record keepers.

2. *Observation of relevant facts.* Observation of the facts is the data-gathering phase as it relates to the problem under study, that is, the learning process. The nature of the data needed will be dictated by the problem. Pertinent information may lie within the records and accounts of the business.

3. *Analysis and specification of alternatives.* The tools of analysis are primarily those of economic principles and budgeting procedures, including mathematical programming. Comparison is fundamental to any decision-making activity.

*For a good discussion of the management process see Warren H. Vincent, ed., *Economics and Management in Agriculture,* Ch. 2, Prentice-Hall, New York, 1962.

4. *Choice of alternatives.* Most decisions are of an either/or type and require measurement involving some welfare criteria. The benefits and consequences must be weighed for each problem decision.
5. *Action and supervision.* The manager must be properly motivated before any action takes place. Many intelligent and clever people fail because they are unable or unwilling to make decisions and take action.
6. *Evaluation and responsibility.* A problem is never really solved until the results of decisions have been evaluated. This may not come until a year or so after decisions have been put into practice. The results generally will show up as changes in production, income, net worth, or some other section of the farmer's record-keeping set. This is analyzed by comparing the records before and after the change as well as comparing the records after the change with the budget projections used to analyze the alternatives. If such analyses are not made, the farmer may never know how effective a manager he is nor what further actions are needed. Responsibility without evaluation is dangerous to the life of any business.

The place of records and accounts in the decision-making process should become more apparent as you move through this chapter.

Records and accounts are historical by definition; they predict the future only in terms of the past. If the farmer is trying to anticipate next fall's cattle prices, he needs to check outlook reports. If he is adding a new enterprise or new machine, he must go to a source where information is available. However, accounts can be useful in illuminating the ailing as well as the healthy segments of the business and, in some cases, in prescribing the appropriate medication. Even though they do not predict the future, they may serve as useful guides. For example, the soil characteristics, past cropping systems, fertilization practices, and past production of a field may be the best available guides to crop yield expectations.

RECORD-KEEPING OBJECTIVES

Major uses of farm records and accounts may be summarized as follows:

1. *Management tool.* Farm records allow the farmer to measure his efficiency in using the factors of production—land, buildings, machinery, labor, etc.—and in producing agricultural products for sale at a profit. Hopkins and Heady have said, "The farmer is an engineer and biologist. . . . As an engineer, he must measure the use of input-output or yield quantities on his farm. As a biologist, he should know what yields he is obtaining under different crop treatments and methods and how plants and animals respond to different treatments. He needs to know how each class of livestock, or each method of handling affects rates of gains, feed requirements, and other basic quantities. This information is required if he is to have even elementary understanding of his current farming operations. It is needed even more if he hopes to improve management and increase profits."[*] Financial accounts measure the farmer's success in producing income for family living and business growth. Farm

---

[*]John A. Hopkins and Earl O. Heady, *Farm Records and Accounting,* p. 14. Iowa State Univ. Press, Ames, 1962.

records and accounts furnish needed information for business analyses and farm planning for effective decision making.

2. *Preparation of income tax reports.* "All taxpayers, including farmers, must keep records which will enable them to accurately prepare an income tax return, and which will permit the Internal Revenue Service to determine whether the law has been correctly applied."[*] While most farmers see the need of records for tax reporting, a considerably smaller number recognize their need for tax management. Experience suggests that farmers who keep poor records pay more taxes rather than less. The objective of tax management is to maximize the after-tax income, not necessarily to minimize the amount of taxes paid.

3. *Basis for credit.* According to the American Bankers Association, "The banker who has records of the borrower's business is able to compare the borrower's past performance against standards for the area. These records also become a basis for projecting and evaluating the future profitability and loan repayment capacity of the business. Records, properly and accurately kept, provide the banker with the financial information needed for prompt handling of credit requests."[†] Records furnish information for planning the credit needs and repayment schedule of the farmer as well as for determining his repayment ability, and they provide evidence of security for production loans and security interest agreements.

4. *Additional uses.* Records and accounts also provide the basis for farm lease arrangements and other contracts, farm insurance programs, and participation in government programs.

These objectives were summarized by Mueller[‡] as functions served by farm accounting systems for individual farmers:

1. Control of financial affairs
   a. Record of bills paid, income received
   b. Accounts payable, accounts receivable
   c. Inventory control
   d. Partnerships, profit-sharing agreements, landlord-tenant settlements, farm corporations
2. Legal and institutional requirements
   a. Income tax: capital gains, investment credit, and investment credit recapture
   b. Social security: self-employed and employee accounts
   c. Historical records: estate settlement, cost basis of real property, ASCS programs
   d. Insurance: coverage, damage claims, and evidence of losses
3. Farm business analysis
   a. Total farm business: trend and comparative analysis and detecting strong and weak points in organization and management performance
   b. Enterprise analysis

---

[*]Internal Revenue Service, Farmer's Tax Guide, annually.

[†]Agricultural Committee, Farm Credit Analysis Handbook, p. I-1, American Bankers Assoc., 1965.

[‡]A. G. Mueller, Application of EDP to Farm Accounting and Farm Management, Ill. Bankers Agr. Credit Conf., 1966.

    c. Lease evaluation
    d. Financial position of business (balance sheet)
4. Basis for planning and budgeting
    a.Information provided by records:
       (1) Basic profit and loss statement on farm unit
       (2) Selected input-output relationships
       (3) Inventory of physical and financial resources available
       (4) Management performance of operator
    b. Applications of record data in planning:
       (1) Projected production and operating plans
       (2) Alternative resource and product combinations compared with existing unit
       (3) Projected financial and cash-flow requirements, credit requirements, and repayment schedules.

COMPONENTS OF A FARM RECORDS AND ACCOUNTING SYSTEM

    This discussion is not exhaustive in its treatment of farm records and accounts; however, major components useful to farm accountants are discussed. A farmer's imagination can carry him beyond this treatment. The accounts, records, and analyses discussed here are:

1. *Asset and liability account.* This is a physical and financial account of all farm resources (assets) and the claims against those resources (liabilities). The proper ordering of the assets and liabilities account will provide the net worth statement or balance sheet of the business. The net worth statement is an account of the farmer's financial position at any point in time.
2. *Receipt and expense account.* This account of financial flows into (receipts) and out of (expenses) the business over a period of time, usually one year, may include both cash and noncash transactions. Subtracting expenses from receipts gives net farm income, which measures the profitability of operating the business, i.e., the return to the operator for his labor, management, and capital.
3. *Capital account.* This is a purchase record of capital assets and improvements, which cannot be debited fully as expenses in the year purchased, and a sales record of similar items. Purchases and sales of capital items will generally affect the depreciation schedule of the business. These purchases and sales directly affect the asset account and generally result in additions to the receipt and expense account.
4. *Credit account.* This record of farm liabilities includes recording new loans as well as keeping track of principal and interest payments and tabulating unpaid principal balances on existing loans.
5. *Production and statistical records.* These records relate to the production of crop and livestock enterprises on the farm and the resources used. Labor, feed, crop fertilization, and crop yield records are examples.
6. *The farm business analysis.* The inventory and the receipt and expense accounts are combined with production records to probe for strong and weak areas within the business. These analyses are commonly called efficiency measures. The information they provide often is useful in identifying problems and directing future farm

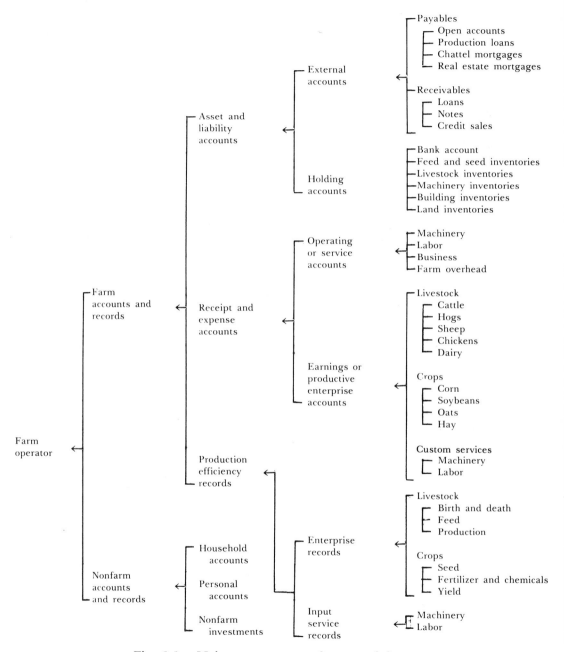

Fig. 1.1. Major components of a set of farm records and accounts.

management decisions. Growth and progress can be measured over time by a comparison between years.

7. *Enterprise records and accounts.* All information that can be recorded for the total farm can be kept for individual enterprises. Often only minor additional records are required, over those necessary for the total business, to make rather detailed enterprise analyses. These often are referred to as cost accounts.

These various accounts are summarized in Figure 1.1 to show their relation to the farm operator.

THE ACCOUNTING PERIOD

The accounting period is usually for one year and typically corresponds to the calendar year. Although this is the common period, it is not necessarily the best for all farm businesses. For some farmers it may be more useful to keep records on a fiscal year basis. A fiscal year is any year having a beginning date other than January 1. For example, the month selected may correspond to the beginning of a lease arrangement, such as March 1. Most government agencies operate on a fiscal year beginning July 1.

The production period should be considered in selecting an accounting period. It may be useful to adopt an accounting period that corresponds to the flow of receipts and expenses from the major enterprises. A broiler producer may wish to balance his accounts after each batch of broilers is sold. Some cropping seasons end in midsummer, and this might be the best time to balance accounts. The time period selected should be the one most meaningful in analyzing the activities of the business. An accounting period can be looked upon as the time between one complete business summary and the next, as illustrated in Figure 1.2.

The net worth statement gives the financial position of the business at a point in time, in this case at the beginning of the accounting period and again at the end. Between the beginning and ending points have been many changes brought about by production, purchases and sales, loan payments and new loans, etc. These activities are recorded in the accounts and records shown in Figure 1.2 and defined earlier. Whereas the net worth statement is static in concept, the receipt and expense accounts, production records, and capital and credit accounts are dynamic and explain the changes reflected by a comparison of net worth statements.

As a business grows, it is necessary to maintain greater control over its financial

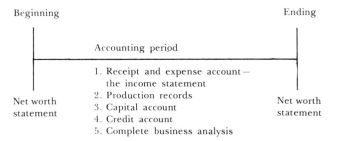

Fig. 1.2. The accounting period.

transactions. Annual summaries may no longer be sufficient to guide the expanded activities of the business. Quarterly and sometimes monthly summaries of major or key segments of the business are required. Through the use of electronic data processing, these can be made available as frequently as may be useful.

Once the accounting period has been selected, it is important for the time to be uniform between accounting periods. If it is not, the business summaries and analyses will not be of the same duration nor will they cover the same periods in the annual business cycle.

A tabulation of net income just prior to the end of the calendar or fiscal year can also be very useful; a major use would be for tax management purposes. For example, if a farmer knows approximately what his taxable income will be, he may be able to plan sales and purchases to level it over time and thus reduce the total taxes he must pay. For this reason a farmer may wish to make a tabulation 6-8 weeks before the end of the taxable year. This would allow him ample time to plan sales and purchases in an effort to influence his tax liability for that year. In many cases if a farmer delays income, he is merely shifting it to another year. The farmer using a cash-basis tax report has more flexibility in influencing the current year's income than the farmer using one on an accrual basis. However, both individuals would have some flexibility in adjusting income for the current year. (The details of income tax management will be discussed in Chapter 10.)

THE ACCOUNTING ENTITY

It is important to identify which financial transactions properly belong to the farm business. First, there are personal or family items. Examples of these transactions may include food, clothing, depreciation and repair on the family dwelling, family auto expense, family life and health insurance, etc. Second, other business transactions, although genuine, may not be part of the farm business being studied. Examples may include investment in stocks and bonds, a wage job off the farm, interest in a nonfarm business such as a cooperative elevator, or even another separate and distinct farm business. These various transactions are shown in Figure 1.3.

The area to which each account applies is called an *accounting entity*. The home and family form a separate accounting entity from the farm. Likewise other business ventures each form separate business entities. A farmer should keep detailed records of each of these three levels of activities. Since there frequently may be transfers of funds and resources among the three levels, it is not always easy to identify where each item

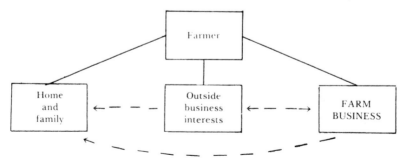

Fig. 1.3. Divisions and relations of farm and family record-keeping activities.

appropriately belongs. For example, what part of the house or auto are rightfully business expenses and what part family? Are the farmer's clothes a personal expense or a business expense?

One of the major purposes of nearly all business ventures is to produce income for family consumption, currently and in the future. To accomplish this objective, financial flows from the business to the household must take place; however, conflicts may exist. If the family demands a new television set, a piano, or a college education, this may mean cutting back on a productive farm investment. Similar flows may exist with other business interests. A farmer may take a job off the farm in order to make mortgage payments. On the other hand, he may make other business investments from the profits that have accrued from his farm business.

It is not our purpose at this point to identify which type of transaction should be recorded where. The appropriate pattern for entering the various transactions should become clearer as the various types of records and kinds of transactions are discussed. If we realize that several levels of record keeping may exist and learn to identify them, we will clarify many of the problems that trouble beginning accountants.

Even though each of these activity levels should be kept separate, it may be useful to bring them together into a composite account. How else could the farm operator determine what kind of a total manager he was? Net worth statements can be joined to determine total net worth and to measure capital increases. Income statements likewise can be joined to determine total net income and to balance total income with expenses and investments. Cash-flow statements often include the total of all income and expense transfers, thus determining the financial needs of the business.

The concentration in this discussion will be on the farm business, but the records and procedures discussed are relevant to all other business activities with which the farmer is involved, including those of his family.

ACCOUNTING PRINCIPLES

In order to fully understand and interpret financial reports, it is necessary to have a clear understanding of the rules under which they were compiled. Also, for any accounting system to function properly, it must have some commonly known and relatively fixed points of reference. Listed here are the principles that are the foundation of most accounting systems.

1. *Serviceability*. Financial accounts are developed to serve the needs of management, which may be business oriented or legally required. There is an added cost for every new record or account introduced, which may show up indirectly as less time for the farmer to do other things rather than as a direct outlay of dollars. This is called *opportunity cost* in economics and measures the cost of one activity in terms of what must be given up in other activities to make the change. A new record should not be added unless it has prospective benefit and that benefit is larger than its opportunity cost. The old benefit-cost principle is still valid in decision making.

2. *Objectivity*. Financial reports should reflect only facts supported by objective evidence of complete transactions. These data are verified by supporting documents and business papers such as invoices, contracts, canceled checks, physical counts, freight bills, time tickets, etc. The financial reports will thus be unbiased and can be verified by independent investigators. For most public and many private institutions

it is a legal requirement that financial accounts be audited periodically. The auditor must be able to verify the accounts by reproducing the same results after objectively submitting them to detailed analyses. More important, the users of the reports must be able to rely upon the summary statements for management control.

A suggested filing system will be introduced in Chapter 13 where procedures will be outlined for filing financial documents and business papers.

3. *Disclosure.* Full supplementation of the pure numerical recordings and tabulations with explanatory comments and footnotes should be made. This may include contracts, pending lawsuits, details of credit transactions, lease commitments, mortgages, changes in accounting procedures, etc.

4. *The going concern.* It is assumed that the business entity will continue its activity indefinitely. Thus only the cost of the asset paid by the entity is considered. The business will have continuous use of the assets for the purpose for which they were acquired. Any deviations from this will be completely identified and thoroughly explained.

5. *Materiality principle.* This provides a practical approach to solving problems and recording details. For example, by definition the cost of any asset with a useful life of over one year should be prorated over the life of the asset. The materiality principle would modify this definition to include only those assets with a purchase price over some specified dollar amount. For the small firm this might be $25 or for the large firm, $100. Determination of this is a matter of experienced personal judgment; chasing a penny accounting error may cost dollars.

6. *Conservatism.* This is the practice of not being overly optimistic. Valuing property, selecting depreciation rates, estimating quantities on hand, etc., are often matters of professional or personal judgment. The principle of conservatism would say to select the more moderate approach. It is better to err on the safe side; however, some feel that accountants have been too conservative and thus have kept management from making wise investment decisions.

7. *Consistency.* Treatment of items in a consistent manner from one accounting period to the next permits reliable comparisons to be made. For example, there are different methods of valuing inventories and depreciating property and if the method changed from one period to the next, business analysis would be difficult and nonproductive.

8. *The stable monetary unit.* This assumes that the general price level remains reasonably constant. Since the primary purpose of accounting is measurement of income and business worth, there must be some stable measuring unit. Thus, if a 5% inflation exists in one year, the accountant does not reduce the value of all new purchases by 5%.

9. *Matching expenses against revenues.* Income in any period may be determined by matching costs against revenues for that period for both cash and noncash items. This method prescribes that the accrual method of accounting is the only truly accurate method of tabulating income for any one period. (Cash and accrual methods of accounting will be discussed in Chapter 5.)

10. *The cost basis of valuation.* The practice of valuing assets at their original verifiable cost assumes that the price agreed upon in the transaction was arrived at by both buyer and seller trying to drive the best possible bargain. Cost includes incidentals such as transportation, installation, title searches, etc. Where there are two alternatives for value, such as in a trade or a gift, the price selected is the one that

represents the more reliable estimate of the fair market value. This principle must be modified for depreciable properties and prepaid expenses.

11. *Realization of revenues*. Revenues are realized when the marketplace transactions increase the owner's equity and should not be recorded until that time. Other possibilities in chronological order are:

a. Signing of a sales contract
b. Purchase of the asset to be sold
c. Production of the asset
d. Appreciation in value of assets held for later sales
e. Delivery of the asset to the customer
f. Receipt of cash from the customer
g. End of a guarantee period on the asset

In summary, an accountant's job is to conservatively, objectively, and with full disclosure record for a uniform time all financial facts, from period to period, on a consistent basis for an accounting entity that is a going concern.

ACCOUNTING: A BRANCH OF ECONOMICS

Economics is the broad field of study that encompasses all man's wealth-getting and wealth-using activities. Accounting is the portion of this field that concentrates on gathering, classifying, and interpreting information for an entity. In this text the entity is the farm unit (manager or operator). Accounting provides information that may be combined with economic theory and analysis techniques for making business decisions. Net income (an accounting term) is similar to profit (an economic term) and both describe a financial position related to wealth. Thus, it is desirable that the accountant be familiar with economic principles and analysis techniques. Only then can the data be prepared in a form and at a time useful to management.

For example, the profit maximizing principle of marginal costs and revenue states that profit will be maximized where output is increased to the point where the added cost is just equal to the added revenue for producing a unit of product. The costs in this case are the business expenses, the returns are the business receipts, and profit is the difference or net income. This net income will become larger as the business expands its activities until the last added unit just pays for itself. Accounting is the only means by which this point can be measured.

2

CAPITAL ACCOUNTS AND DEPRECIATION SCHEDULE

CAPITAL INVESTMENT ACCOUNTS

As defined by economists, capital refers to the productive resources that are the result of man's creation. Excluded are natural resources, labor, and management. However, in accounting and investment, capital takes on expanded and sometimes restrictive meanings. For example, the equity an individual has in his business is often called his capital. The cash that a firm raises to carry on its business is called its capital. A capital balance may refer to a difference in trade between two countries. The Internal Revenue Service (IRS) defines capital assets as only those items held for personal purposes, pleasure, or investment. Thus modifiers are used to designate particular kinds of capital. In this chapter investment capital is used to define particular resources used in the farm business. All the purchases a farmer makes to operate his business could properly be called capital, including feed, seed, fuel, repair items, livestock, machinery, and buildings. Investment capital refers only to items that are not purchased directly for resale purposes, give services to other productive business activities, and have a service life of over one year. Included in this definition would be breeding livestock, machinery, buildings, equipment, and various land improvements such as fences, wells, land terraces, and water drains.

The fact of purchase of investment capital assets does not constitute a business expense. At the moment of purchase the asset still retains its market price or value. The time factor of ownership and use causes the market price to change, and it is this element that constitutes a business expense. Thus the making of a capital investment purchase merely changes the form of capital held. Money capital is traded for livestock, machinery, buildings, or similar investment capital. This is true even if the purchase is on credit. In this case the business firm making the capital sale also is assuming the role of a credit agent. It is loaning money to the buyer to make the purchase; a liability has been created. (The accounting of this kind of transaction will be treated in Chapter

3.) The purchaser is no better or worse off immediately after the purchase whether the purchase was for cash or credit. He makes the purchase with the belief that the new asset will provide a return on his investment higher than from holding his dollars, even if he must pay interest on the money borrowed to make the purchase.

The business expense of holding investment capital (whether used or not) is the change in value that takes place within the accounting period. (For a new purchase this may not be for a full accounting period in the year of purchase.) The method commonly employed to estimate this change in value is depreciation, which is a systemized mathematical approach for estimating the change in market price for most of the investment capital assets here defined. Thus many of the items entering the capital purchase account will be transferred to the depreciation schedule. (Depreciation methods and accounts will be discussed later in this chapter.) The point to be recognized here is that sufficient detail must be kept in the capital investment account to make this transition.

Often a trade is involved in the purchase of a new capital asset. The purchaser turns in a used item and pays a cash difference. A value must be placed on the asset traded; in fact it may have two values — the allowance given by the seller of the new asset and the value the asset had in the accounts of the purchaser. Thus the total purchase price of the new item is the cash difference plus the trade-in allowance; together they constitute the market price of the item purchased. However, from the standpoint of the purchaser the amount of total investment in the new item is the dollars of trade difference plus the value the traded item had in his accounts. In tax accounting, this is termed the *tax basis*.

It is not necessary that the purchased asset and the traded asset be identical or even serve the same general purpose in the business, but for the value of a trade-in item to be counted in tabulating the basis of a new capital asset purchase, it should be a "like item." A cow for a plow would not be like items, but a plow for a tractor would be. If the item traded is not a like item, the trade-in should be considered a sale and the basis of the new investment would be its market price.

Thus the capital asset investment purchase account should provide for recording the following:

1. Date of purchase
2. Description of item purchased
3. Description of item traded
4. Total market purchase price of item purchased
5. Cash difference between the market price and the allowance given for the item traded
6. Cash paid down where dealer financing is involved
7. Value of the item traded in the accounts of the purchaser
8. Tax basis of the item purchased

Such an account is illustrated in Form 2.1. Where a trade-in is not involved, some columns will go unused; but accounts must provide for recording the detail of all transactions, not merely most of them.

Breeding livestock may be an exception and may not be recorded in this account. It would not be improper to record them here, but most farm account books place breeding livestock purchases in the account with feeder livestock purchases. (Livestock purchase accounts are discussed in Chapter 5.) This places all livestock purchases in the same account and thus may be a more convenient reference. However, a major reason for not

Form 2.1. Capital Asset Purchases and Expenditures for the Dunn Farm

Includes the purchase of depreciable machinery and equipment and buildings, improvements, and major repairs to depreciable property. It does not include breeding livestock purchases even though they may be depreciated.

Provision is made for recording items traded and for calculation of the depreciable tax basis. Items are to be transferred from here to the depreciation schedule. Similar categories of items may be grouped within the columns.

	Date	Item purchased or improved	Item traded	Purchase price of new item	Trade-in allow-ance	Cash differ-ence	Cash paid down[a]	Value of item traded[b]	Tax basis[c]
				$	$	$	$	$	$
1	Machinery and Equipment								
2	1-08	truck - 1 ton		4400 00		4400 00	1000 00		4400 00
3	3-27	tractor - JD	tractor - IH	6100 00	2100 00	4000 00	500 00	1600 00	5600 00
4	9-20	plow JD-5	plow IH-4	1300 00	100 00	1200 00	1200 00	200 00	1400 00
5									
16	Buildings and Improvements								
17	8-8	Reroof shed		400 00		400 00	400 00		400 00
18	8-30	Grain bin (3000 bu)		1000 00		1000 00	1000 00		1000 00
19	11-23	Hog feeders (3)		300 00		300 00	300 00		300 00
20									
32									
33	Column Totals			xxx	xxx	xxx	$ 4400 00	xxx	13,100 00

a Cash paid above any dealer loans obtained to finance purchase.

b See the depreciation account for undepreciated value.

c Cash difference plus value of item traded.

including breeding livestock in this section is that they are not always placed in the depreciation schedule. The change in value of breeding livestock may be measured more directly in terms of market quotations or some other method of valuation. (The valuation of farm assets will be discussed in more detail in Chapter 4.) Regardless of where it is recorded, the essential transaction detail is the same. Also, the net worth and income statements are similarly affected regardless of where they are recorded. This will become clearer as other accounts are discussed.

If the dealer finances part or all of the sale, a liability has been created, as already mentioned, and the cash difference and the cash paid down will not be the same. The cash difference is the difference between the purchase (market) price and the trade-in allowance, whereas the cash paid down is the cash payment made at the time of purchase. If the dealer is not financing the purchase, the cash difference and the cash paid down will be the same. This does not mean that some other source has not financed the purchase. Dealer-financed purchases should be recorded as loans in the credit accounts. Neither the purchase price nor the cash difference reflects the cash flow out of the business at the time of purchase. The "cash paid down column" indicates the total cash drain on the business as a result of the new purchase. This is a useful figure when tabulating the balance of cash coming an in and going out of the business. (This concept will be explored more in Chapter 5.)

Capital asset sales also should be recorded. Assets disposed of through trade for a like item already have been discussed. Direct sales of investment capital and credit received through trade where the item traded was unlike the item purchased should be recorded in this account. The items normally recorded in the capital asset sales account are depreciable properties. For the reasons already discussed in reference to the capital assets purchase account, breeding livestock usually are not recorded in this account even if they are on the depreciation schedule. Normally they are recorded in the livestock sales account as the sale of breeding or butcher livestock. However, it would not be improper to show them in the capital account.

If the value received in sale or trade for unlike items is different from the undepreciated balance or adjusted basis for depreciable properties, a gain or loss has been sustained. If the sale price is larger than the account book value (usually the undepreciated balance) there is a gain, and if it is less there is a loss. For depreciable properties this may be conceived as a depreciation adjustment; i.e., if the depreciation schedule truly reflected market values, the amount received in sale would be the same as the undepreciated balance. However, neither depreciation methods nor market prices are that accurate or reliable, but there is a bit of logic in the comparison. In any case it represents an income adjustment, and the difference should be recorded in the income statement. (How this is handled will be discussed in Chapter 5.)

Items to be recorded in the capital asset sales account should include:

1. Date of sale
2. Description of item sold
3. Amount received through sale
4. Cash received from sale if it is different from the sales amount
5. Value that the item sold had in accounts of the seller at time of sale
6. Gain or loss through sale

A capital asset sales account is illustrated in Form 2.2. Again, the amount of total cash received through sale is of cash-flow interest.

Form 2.2. Capital Asset Sales for Cash

Includes cash sales (like items trade involved) of depreciable properties and sales of nondepreciable assets not covered elsewhere, such as small tools and equipment. A profit is shown where the sale is for more than the undepreciated balance at the beginning of the year and vice versa.

	Date	Description of item sold	Sales amount	Cash received	Value of item sold[a]	Gain or (loss)
1	4-30	Manure spreader	$	$ 200 00	$ 100 00	$ 100 00
2	11-23	Farrowing stalls (2)		50 00	75 00	(25 00)
3						
4						
5						
31						
32						
33	Column Totals		$	$ 250 00	xxx	$ 75 00

[a]See the depreciation account for undepreciated value of item sold.

DEPRECIATION

Depreciation is a systematic method for arriving at the value of a working asset based upon its cost as well as a method of prorating the cost of a working asset over its productive life. Working assets give productive service to the business over a period longer than one year and in so doing become less valuable. A more restrictive definition of working assets will be given later. Assets commonly found in the depreciation schedule are: tractors, plows, planters, and other similar machinery; silo unloaders, mechanical feeders, barn heaters, and other equipment; fences, wells, drains, and other improvements; barns, silos, sheds, tenant houses, and other buildings; and purchased breeding livestock. These are the assets that are normally recorded in the capital asset purchase account discussed earlier in this chapter. Land and some permanent improvements such as clearing of trees and bushes, leveling, and water ponds are not usually considered depreciable. Thus the major purposes for depreciating an asset are:

1. To estimate current value
2. To calculate expense
3. To adjust taxable income

The last purpose will not be explored in detail in this chapter but should be part of a farmer's thinking whenever he purchases a depreciable asset or selects a method of depreciation. By selecting different methods of depreciation, farmers may be able to adjust their business expenses. The purpose of this is to level out taxable income over time and thus reduce the total tax payment. (Tax management is discussed in Chapter 10.) These three purposes should be kept in mind in the discussion that follows.

The major causes associated with depreciation are:

1. *Wear and tear.* This is caused by the wearing out of an item through use.
2. *Obsolescence.* This is a time consideration and relates to new technical developments that render the old machines less useful or desirable as compared to newer ones.
3. *Deterioration.* This is change in value due to elements of nature. For machinery this might be rust; for buildings, decadence; for livestock, aging.

Causes 2 and 3 are changes that take place even if the item is not productively used. Thus they are fixed-cost elements, which the farmer has after purchase whether he uses the item or not. Wear and tear is a variable cost, one which the farmer has only when he uses the item. Since fixed costs are usually greater and more significant than variable costs, depreciation is generally considered a fixed cost.

Many depreciable assets have residual value after they are no longer useful in the business. This is commonly called *salvage value.* This may be an item's scrap value, its market value when the farmer plans to dispose of it, or its value when it ceases to depreciate. For most machinery and buildings salvage value is generally scrap value. For breeding livestock it is normally anticipated cull market value. *The amount to be depreciated,* then, is the difference between an asset's cost (or basis as defined later in this chapter by the IRS) and its salvage value.

Before calculating the annual depreciation, it is necessary to estimate the asset's useful or depreciable life. This is mainly an individual matter and relates to the farmer's own experiences; however, the experiences of other farmers are sometimes helpful. A

TABLE 2.1. Expected service life for various farm machine units

Machine	Annual use	Useful life (years)	Machine	Annual use	Useful life (years)
Powered Machines			*Cultivators, Weeders, Sprayers*		
Automobile	6,000-8,000 mi	10	Cultivator:		
	8,000-10,000 mi	9	4-row	100-200 A	12
	10,000-12,000 mi	8		200-400 A	10
Truck:					
Pickup	4,000-6,000 mi	13	Field	. . .	14
	6,000-8,000 mi	12			
	8,000-10,000 mi	11	Rotary hoe	. . .	12
1½-2 ton	2,000-4,000 mi	15	Sprayer, tractor-		
	4,000-6,000 mi	13	mounted	. . .	10
	6,000-8,000 mi	12			
			Harvest Machines		
Tractor:			Corn picker, mounted and pull:		
Wheel	200-400 hr	14	2-row	85 A avg.	10
	400-600 hr	12			
	600-800 hr	11	Self-propelled	140 A avg.	10
	800-1,000 hr	10			
			Combine:		
Crawler	400-600 hr	14	Pull type	150-200 hr	10
	600-800 hr	13			
			Self-propelled, 12 ft	200-250 hr	10
Seedbed Preparation Machines				250-300 hr	8
Plow, one-way:					
3-bottom	100-150 A	12	Mower:		
	150-200 A	10	Rotary	. . .	12
4-bottom	150-200 A	12	Sickle bar	. . .	12
	200-300 A	10			
			Side rake	. . .	12
Disk harrow, 8-10 ft	100-200 A	12			
	200-300 A	11	Hay baler	. . .	8
	300-400 A	10			
			Field chopper	. . .	8
Spike harrow	. . .	12			
			Ensilage blower	. . .	10
Lister	. . .	12	*Other Farm Machines*		
Roller	. . .	15	Wagon and trailer	. . .	12
			Commercial fertilizer		
Deep tillage			spreader	. . .	8
machinery	. . .	12			
			Manure spreader	. . .	14
Planters					
Grain drill, 8-10 ft	50-100 A	14	Post hole digger	. . .	10
	100-150 A	12			
	150-200 A	10	Tractor scoop and		
			blade	. . .	12
Corn planter, 4-row	100-200 A	12	Grinder and		
	200-300 A	10	hammer mill	. . .	10

Source: Summarized from various studies conducted by agricultural experiment stations between 1960 and 1970.

TABLE 2.2. Guideline lives for farm assets

Asset	Life (years)
Farm buildings	25
Land improvements	20
Machinery and equipment (including fences)	10
Transportation equipment (commercial):	
Automobiles	3
Trucks (less than 13,000 lb unloaded)	4
Trucks (13,000 lb or more loaded)	6
Trailers	6
Animals:	
Cattle, breeding and dairy	7
Horses, breeding or work	10
Hogs, breeding	3
Sheep and goats, breeding	5
Trees and vines	according to practice

summary of useful lives for farm machinery from various studies conducted by agricultural experiment stations appears in Table 2.1. In 1962 the IRS issued Revenue Procedure 62-21 which established 75 broad classes of guideline lives. These recommended ''lives'' for the major classes of assets used by farmers are shown in Table 2.2.

A comparison of Table 2.1 with Table 2.2 reveals that the guideline lives are shorter than farmer experiences. One explanation for this is that the farm surveys tended to obtain estimated lives for the machines then in use. Machines taken out of use at earlier ages would not have been included in the average. Also, machines discarded by one farmer tend to show up on some other farm where the annual use is smaller. In addition, major repair jobs tend to lengthen the lives of many machines; thus the lives shown in Table 2.1 probably are on the long side.

The guideline lives, if used, are to be applied to all assets in a particular class and could be modified within limits according to the experiences of the farmer using them. To use these lives, farmers must meet other requirements of the law which will not be discussed here. The 1971 Revenue Act replaced the 1962 procedure with the Asset Depreciation Range System, which established class lives. These lives were basically the old guideline lives, give or take 20%. Again, it was necessary to meet specific qualifications to use these lives. Since these procedures are subject to change, they will not be discussed here. The reader is referred to the current Farmer's Tax Guide and other publications of the IRS. Old procedures have not been replaced and it is doubtful that they will be. The new procedures are simplifying conventions for some farmers and may give them certain advantages. A farmer must study the new procedures and test them against his current practices to determine if a change is worthwhile. The guideline lives are presented here to illustrate depreciable property lives that may be useful to farmers who have not had sufficient experience of their own.

The three methods of depreciation most commonly used are:

1. Straight-line method
2. Declining-balance method
3. Sum-of-the-years-digits method

Even though only three methods will be discussed, any method that will give a reasonable depreciation allowance consistent with the purposes previously discussed and within legal limits is acceptable. For illustrative purposes when discussing these methods, assume the following situation:

1. Truck purchased new in 1971 for $4,400
2. Ten-year life
3. $400 salvage value
4. A full year's depreciation the first year

The *straight-line method* (SL) is the old standby used by most farmers. Under this method the cost or other basis of the property less its estimated salvage value is deducted in equal amounts over its estimated useful life.

$$\frac{\text{Annual}}{\text{depreciation}} = \frac{\text{cost (or basis) minus salvage value}}{\text{estimated useful life}}$$

The annual depreciation deducted each year is a constant.

> *EXAMPLE 2.1*
>
> Using the straight-line method, the 1971 depreciation for the truck would be:
>
> $$(\$4,400 - \$400)/10 = \$400$$
>
> The value of the truck at the end of the first year is $4,400 - $400 = $4,000, at the end of the second year its value will be $4,000 - $400 = $3,600, at the end of the third year $3,200, etc., until at the end of the tenth year only the $400 salvage value remains. If the tractor were purchased at midyear, say April 1, the first-year depreciation would be (9/12) x $400, or $300. Each year thereafter the depreciation would be $400 until the last part of a year when it would be $100($400 - $300).

The *declining-balance method* (DB) is a fast write-off method. The largest depreciation deduction is taken the first year and becomes smaller in following years. Under this method a constant percentage of the remaining undepreciated balance is taken as depreciation each year:

$$\frac{\text{Annual}}{\text{depreciation}} = \left(\begin{array}{c}\text{constant} \\ \text{percent}\end{array}\right) \times \left(\begin{array}{c}\text{undepreciated balance} \\ \text{at beginning of year}\end{array}\right)$$

Under this method no salvage value is deducted from the purchase price or basis but is held as the minimal value to which the item can be reduced. The IRS has placed these restrictions on the use of this method:

1. The asset's useful life must be at least three years.
2. A maximum of twice the *rate* under the straight-line method can be applied for new personal properties, and 1½ times the rate for new depreciable real property, primarily buildings and their structural components, and for used personal property. Used depreciable real farm property must be depreciated by the straight-line method.

(Prior to the Tax Reform Act of 1969 depreciable real property was under the same rules as depreciable personal property.)

The rate of depreciation under the straight-line method is calculated by dividing the length of life into 100%. In Example 2.1 the rate would be 10% (100% ÷ 10 years). Thus the maximum rate under the declining-balance method for the example problem would be 20% (2 x 10%). For this reason the declining-balance method is frequently referred to as the double-declining balance method or 200% declining-balance method. Twice the rate under the straight-line method is the same as dividing the length of life into 200%. One and one-half times the straight-line rate referred to as the 150% declining-balance method. As a maximum, under the 200% declining-balance method, 66⅔% of the purchase price of a new asset could be claimed as depreciation the first year (200 ÷ 3).

EXAMPLE 2.2

Applying the declining-balance method to the example problem, the following depreciations and values are obtained:

First-year depreciation = 0.20 x $4,400 = $880
End-of-year value = $4,400 — $880 = $3,520
Second year depreciation = 0.20 x $3,520 = $704
End-of-year value = $3,520 — $704 = $2,716

If, instead of a full year, a part of year was involved, say an April 1 purchase, these computations would be:

First-year depreciation = 0.20 x $4,400 x (9/12) = $660
End-of-year value = $4,400 — $660 = $3,740
Second-year depreciation = 0.20 x $3,740 = $748
End-of-year value = $3,740 — $748 = $2,992

The *sum-of-the-years-digits method* (SYD) falls between the other two methods previously discussed. It is calculated as follows:

$$\text{Annual depreciation} = (\text{cost} - \text{salvage value}) \times \frac{\text{remaining life in years (beginning of year)}}{\text{sum of digits (total years of life)}}$$

The same IRS restrictions apply to the sum-of-the-years digits method as to the declining-balance method.

EXAMPLE 2.3

Applying the sum-of-the-years-digits method to the example problem, the following depreciations and values are obtained:

First-year depreciation

$$= (\$4,400 - \$400) \times \left(\frac{10}{10 + 9 + 8 + 7 + 6 + 5 + 4 + 3 + 2 + 1} \right) = \$727$$

Deciding which method to use is troublesome if the purposes of the valuation are not emphasized. Keeping in mind the purposes of establishing a realistic market value as well as a business expense, the following guidelines might be useful in deciding which method to apply:

1. Powered machinery (trucks, tractors, combines, silo unloaders, etc.) normally change in value faster in the early years than in the later years of their life, so the declining-balance method usually fits these items best.
2. Nonpowered machinery (plows, planters, manure loaders, wagons, etc.) normally change in value about equally throughout their life, so the straight-line method usually fits these items best.
3. Farm buildings do not have an established market value, so the depreciation method selected should be related to the services provided. Usually the straight-line method approximates these.
4. New untried machines do not have an established length of service life. For these it may be wise to select a faster write-off method — the declining-balance or sum-of-the-years-digits method.
5. Small tools and equipment (hammers, wrenches, shovels, feed pans, buckets, etc.) whose purchase price is relatively small and whose service life is undetermined should be inventoried but probably not depreciated. Normally, annual replacement costs are about equal to the depreciation. It seems more realistic to treat the replacement cost as a current expense and hold the inventory value nearly constant.

Income tax and other considerations are relevant in selecting a depreciation method. Remember, however, that the method selected may not affect the total farm depreciation. Only 100% of the depreciable value of an asset can be charged. If the amount of depreciation is greater in early years, it is less in later years. If equal amounts of new depreciable properties are added to the total inventory each year, the selection of a method makes little difference. However, farmers tend to group their purchases of new properties according to fat and lean years in the business, and the selection of a method becomes important as an income leveler and tax adjuster.

Many farmers use the straight-line method of depreciation because of its simplicity. With other methods an annual calculation of allowable depreciation is necessary. Farmers often feel it is too difficult and time consuming to use the declining-balance or sum-of-the-years-digits method. For the same reasons many tax advisers encourage farmers to use the straight-line method. If a farmer purchases about the same amount of depreciable property each year, the end result is not greatly different over time with any method. However, from a tax management standpoint a farmer should utilize the method that gives him the greatest potential advantage.

One other consideration in selecting a method is the cost of using a depreciable asset in the business. If the services provided by the asset do not decline with age and a fast write-off method is selected, this gives a distorted business expense as related to the productivity or service rendered. One other aspect of this relates to repair. The repair costs usually increase with age of the property. If the straight-line method is used, the larger business cost — depreciation plus repair — would come in the later years of life when the service rendered may be less. The service and repair functions should be considered when selecting a depreciation method.

Buildings no longer serving the function for which they were originally constructed present a special problem with respect to depreciation and valuation. Barns constructed to house draft horses are now being used for cattle, cattle sheds for machinery, chicken houses for hogs, etc. From a practical standpoint the original structure can be depreciated according to original specifications as long as the first owner retains it. Of course, major changes to the original structure would enter the depreciation schedule as separate entries and be depreciated according to the improvement cost, length of life, etc. This would be the same regardless of the building's intended use. The difficulty in valuing buildings comes mainly when the building changes ownership. In this case the value of the old building is in terms of what the new owner intends to do with it. The value of the old chicken house that is now being used to farrow sows is in terms of a farrowing house. Thus, a building's value is what it would cost to replace it with a building that would provide the same intended use, depreciated to its present condition. *Replacement cost minus depreciation* may be an appropriate method to use when giving value to a newly purchased building.

One of the difficult problems facing the farm accountant is how to handle major repair items. Should a repair be counted as an investment and be depreciated over a period of years, or should the total repair cost be counted as a business expense in the year made? Three guideline questions might be asked:

1. Does the repair increase the life of the asset being repaired?
2. Does the repair cost have a residual value that can be recovered through sale at the end of the current accounting period?
3. How large is the repair expenditure relative to the size of the business?

If the repair increases the life of a depreciable asset, it probably should be capitalized and depreciated. Likewise, if the repair cost increases the market value of the asset even though it does not increase the asset's life, it probably should be capitalized and depreciated. The third item is not so easy to answer; it relates to the materiality principle discussed in Chapter 1. This principle suggests a practical approach. A small expenditure may not justify the accounting cost of adding it to the investment account and depreciation schedule. For a small farmer, retiring the tractor may be a major investment, whereas for a large farmer this may be an annual event and thus be considered a routine expense.

A related problem is whether the repair investment should be added to the value of the asset repaired or treated as a separate item in the accounts. Again, there is no set answer but if the repair is separable from the asset being repaired, it probably should be treated separately in the accounts. If the repair item becomes an integral part of the asset repaired, the depreciation schedule of the asset should be adjusted accordingly. This same reasoning can be applied to new purchases. It probably is better to treat component parts of a machine or building as separate investments, and thus separate depreciation entries, if they can be separated and there is a chance they will be traded separately. An example of this is a combine with a separate corn head and platform. Another related reason for separate entries is different wear-out lives.

The IRS has given the following restrictions and modifications which should be considered when depreciating properties:

1. The salvage value of depreciable property other than livestock with useful lives of three or more years may be reduced by an amount up to 10% of the basis of the property.

2. Once a depreciation method for an item is selected, it cannot be changed except when changing to the straight-line method from faster write-off methods.
3. Where like item exchanges are made, any gains or losses on the item traded must be reflected in the basis of the item purchased.
4. Tangible personal property purchased new or used with a useful life of six or more years is eligible for a special write-off in the first year it is owned. This is called "additional first-year depreciation." The maximum amount is 20%.

The IRS defines *basis* to mean "the cost or other original basis assigned to the property when you first acquire it. . . . The basis of property purchased for money is ordinarily its cost. . . . If property is acquired in a taxable trade, your basis of the property acquired is its fair market value. . . . If property held for productive use in a . . . business . . . is traded for property of a like kind to be held for a like purpose, the gain, if any, is included in income only to the extent of the 'boot' received (cash or unlike property). Losses resulting from trades in kind are not recognized even though unlike property or money is received in addition to like property. . . . In trades of this sort, the basis of the property acquired is the same as the adjusted basis of the property traded, minus the amount of money received, plus the amount of gain included in income and plus the cash paid." The *adjusted basis* is defined to mean the "basis increased by any improvements or alterations" which would raise its value *or* "decreased by depreciation, depletion or amortization deductions, or by deductions for losses such as casualty loss" which would lower its value.* In summary, the adjusted basis is normally the undepreciated balance or end-of-year value of a depreciable asset. The basis is normally the cost or market price of the asset if purchased outright or obtained in an unlike item trade *or,* when a like item is traded, the cash difference or boot plus the adjusted basis (undepreciated balance) of the item traded.

EXAMPLE 2.5
Calculation of basis where a gain is received. Given:

Cash difference (boot) paid	$ 700
Trade-in allowance on item traded	300
Market price of item purchased	$1,000
Adjusted basis (undepreciated balance) of item traded	$ 200
Gain on item traded ($300 − $200)	100
Basis of item purchased ($1,000 − $100)* or ($700 + $200)†	900

*Market price plus gain on item traded.
†Undepreciated balance of item traded plus cash differeence.

Thus it can be seen that the gain received on the item traded was used to reduce the cost of the item purchased. It may be useful to consider the above in terms of recovered and unrecovered costs. Purchase costs of depreciable assets are normally recovered through depreciation and/or sale.

Unrecovered cost of item purchased	$1,000
Unrecovered cost of item traded	200
Total unrecovered cost	$1,200
Recovered cost on item traded	300
Net unrecovered cost	$ 900

*Internal Revenue Service, Farmer's Tax Guide, 1974.

The newly purchased item is entered on the depreciation schedule at $900. The closing inventory reflects a $200 loss over the beginning inventory with regard to the item traded. No taxable gain is realized.

EXAMPLE 2.6
Calculation of basis where a loss is sustained. Given:

Cash difference (boot) paid	$ 700
Trade-in allowance on item traded	300
Market price of item purchased	$1,000
Adjusted basis (undepreciated balance) of item traded	$ 400
Loss on item traded ($300 − $400)	−100
Basis of item purchased ($700 + $400)	1,100

From an unrecovered and recovered cost standpoint:

Unrecovered cost of item purchased	$1,000
Unrecovered cost of item traded	400
Total unrecovered costs	$1,400
Recovered costs on item traded	300
Net unrecovered costs	$1,100

From an accounting standpoint, whether trades are treated as unlike or like exchanges makes little difference to the long-run net worth or income statements, disregarding tax considerations. Gains or losses on items traded are realized over a period of years (the life of the asset purchased) rather than in the year the transaction was made. Realization is through depreciation adjustments. The intent of the IRS is to encourage more realistic depreciation schedules. This involves both the asset's life and the depreciation method and ties them more closely to market values.

The "additional first-year depreciation allowance" was added to encourage farmers and other businessmen to buy more investment property as a stimulus to the general economy. Regular depreciation may be taken after the additional first-year allowance has been taken. The maximum amount is 20% of the cost (without reduction for salvage) of qualifying property not in excess of $10,000 if a separate return is filed or $20,000 if a joint return is filed. This amount may be claimed if the property was owned any part of the year and is in addition to regular depreciation allowances.

When figuring first-year depreciation, the full purchase price cannot be used if a trade-in is involved. *Only the cash difference "new money" is multiplied by the 20%.* If the same example as above is used and the asset is purchased outright, the special depreciation and the first half-year of ordinary depreciation are figured as in Example 2.7.*

EXAMPLE 2.7
Calculation of first-year depreciation allowance for a new purchase with no trade involved (10-year life, straight-line depreciation, purchased midyear). Given:

---

*Internal Revenue Service, Farmer's Tax Guide, 1974.

Purchase price of item (tractor) $4,400
Salvage value   400
     Depreciable balance $4,000
Additional first-year depreciation allowance ($4,400 x 20% = $880) $ 880
Regular depreciation allowable for the first year
[$4,000 − $880 = $3,120; $3,120 ÷ 10 = $312; $312 x (6/12) = $156]
Total depreciation the first year $1,036
Value of item at end of year ($4,400 − $1,036 = $3,364) $3,364

EXAMPLE 2.8
    Calculation of first-year depreciation allowance for a new purchase where a trade is involved. Given:

Purchase price of item (tractor) $4,400
Trade-in allowed on old item   500
     Cash difference paid $3,900
Additional first-year depreciation ($3,900 x 20% = $780) $ 780

Aside from the income averaging aspects of the additional first-year depreciation allowance and fast write-off methods of depreciation there is also an investment advantage. If the tax money saved in the early years is invested and this investment produces a positive return, income accrues to the operator over the life of the asset on the added investment. This is illustrated in Example 2.9, using the 20% first-year allowance as an example. This assumes a larger new investment in years when income is high.

EXAMPLE 2.9
    Illustration of the income advantage of using the 20% first-year depreciation allowance.
    Given: Item purchased for $1,000, estimated life 10 years, no salvage value, opportunity interest return 5%, the individual purchasing the item pays taxes at 20% on the margin.

Year:	1	2	3	4	5	6	7	8	9	10
Depreciation schedule:										
Without first-year depreciation allowance ($)*	100	100	100	100	100	100	100	100	100	100
With 20% first-year depreciation allowance ($)†	280	80	80	80	80	80	80	80	80	80
Income tax reductions on the amount of depreciation claimed:										
Without allowance ($)‡	20	20	20	20	20	20	20	20	20	20
With allowance ($)‡	56	16	16	16	16	16	16	16	16	16
Difference ($)	36	−4	−4	−4	−4	−4	−4	−4	−4	−4
Investment advantage:										
Available money ($)§	36	32	28	24	20	16	12	8	4	0
Interest at 5% ($)	1.80	1.60	1.40	1.20	1.00	0.80	0.60	0.40	0.20	0

Present accumulated advantage = $9.00

*Depreciation tabulation: annually = $1,000/10 = $100
†Depreciation tabulations: first-year allowance = $1,000 x .20 = $200; annually = $1,000 $200/10 = $80
‡Annual depreciation times tax rate.
§The first year $36 was saved in tax payments that could be invested. At 5% return this $36 would earn $1.80. The second year the taxes paid would be higher by $4 than if the first-year allowance had not been claimed. But even after paying the higher tax, there was still $32 remaining for investment the second year, etc.

Form 2.3. Machinery and Equipment Depreciation

Item	Date acquired	Purchase price	Tax basis	Salvage value	Depr. balance	Life or rate	Depr. method	20% Depr. allow.	Invest. credit Rate	Invest. credit Amt.	1970 Depr. amt.	Value, end of year	1971 Depr. amt.	Value end of year
1 Tractor - IT	1-8-71	$4400 00	$4400 00	$400 00	$4000 00	20%	DB	$		$	$	$	$ 880 00	$3520 00
2 Tractor - IH	3-16-64	4600 00	4600 00	600 00	4000 00	10 yr	SL				400 00	1700 00	100 00	Traded
3 Tractor - JD	3-27-71	6100 00	5600 00	600 00	5000 00	25%	DB	100					900 00 / 900 00	3900 00
4 Plow - IH 4	9-25-64	900 00	900 00	100 00	800 00	8 yr	SL				100 00	275 00	75 00	Traded
5 Plow - JD 5	9-20-71	1300 00	1400 00	200 00	1200 00	1/36	SD		7%	98			200 00	1200 00
6 Manure spreader	4-30-67	600 00	600 00	100 00	500 00	5 yr	SL				100 00	133 00	33 00	Sold
7 etc.														
8														
9														
10														
11														
12														
13														
14														
15														
16														
17														
18														
19														
20														
21														
22														
32														
33 All other items of machinery and equipment											640 00	6406 00	746 00	7460 00
34 Column Totals										$	$8514 00		$3734 00	$6080 00

In Example 2.9 it is assumed no offsetting smaller depreciations in prior years will nullify the effects of faster write-off. That is, if the same investment were made each year and a faster method of depreciation were used including the additional first-year allowance, the same total depreciation would result as if the straight-line method were used with no additional first-year allowance. Also, it is assumed that the additional income received from investing the interest income (compounding principle) is nearly equal to the discount rate for the present value of the future income. To illustrate this, consider that the $1.80 earnings from the $36 investment the first year were invested at 5% returns the second year. This $0.09 would be added to the $1.60 for a total future return of $1.69. But the discounted present value of this $1.69 at a discount rate of 5% would be $1.61. Thus no discounting was used in the illustration.

The *depreciation schedule* is illustrated in Form 2.3. The items listed are those included in the purchase and sales of capital assets shown in Forms 2.1 and 2.2. You will note that the purchase price and tax basis columns are filled in from the data recorded in the capital accounts. The depreciation tabulations for the first item listed, the truck, are taken from Example 2.2. Other items illustrate the different methods of computing depreciation, handling trades and sales, recording first-year depreciation allowance, and keeping track of investment credit. Investment credit is an incentive payment to businessmen (taken as a reduction in the amount of income tax paid) to encourage them to invest in capital assets as a stimulus to the productivity of the economy. It has existed sporadically since 1961. (Investment credit will be discussed fully in Chapter 10.) The beginning and ending undepreciated investment values for 1971 and the depreciation for 1971 on all machinery items are shown at the bottom of Form 2.3. (These figures will appear in the net worth and income statements illustrated in Chapters 4 and 5 respectively.)

It is good practice to separate the items in the depreciation schedule according to some use or allocation pattern rather than just listing them in date of purchase sequence. The following divisions are suggested as a beginning:

1. Powered machinery that has general use
2. Specialized crop machinery
3. Specialized livestock machinery
4. General-purpose machinery
5. Livestock by kind
6. Buildings and improvements

This type of separation may not be the most convenient for tax purposes, but for analysis purposes it facilitates the allocation of investments and depreciation expenses to the productive activities of the business.

3

CREDIT ACCOUNTS

THE CREDIT ACCOUNT is for keeping track of money borrowed and money loaned. Since farmers normally do not loan money, most of this discussion will concentrate on various aspects of borrowed money. To aid in understanding credit accounts and information obtained from them, credit instruments and interest rates will be discussed.

PROPERTY LOANS

All farm property can be classified into two broad types, real and personal. Real property (real estate) is the land and all property permanently affixed to it, such as buildings on permanent foundations, silos, nonportable fences, wells, underground water systems, terraces, and tile drains. Personal property is all nonreal property and includes portable buildings, movable machinery and equipment, livestock, crops, and supplies. It is not the function performed by the property that makes it real or personal but how permanently it is attached to land.

The primary types of loans used to finance real property purchases are mortgages and contracts. Over short periods unsecured loans (signature notes) may be used, but these are the exception. A real property mortgage is a credit instrument that sets aside certain real property as security for (repayment guarantee) the loan. The provisions not only identify the property used to secure the loan but also establish a priority claim among lenders where second mortgages (third parties) are involved. Mortgages are usually recorded in the courthouse as a public record for any interested party to see. Title to the property (deed) normally is held by the debtor (borrower, mortgagor, or buyer—i.e., the person taking out the loan) and not held as security by the creditor (lender, mortgagee, or seller—i.e., the person making the loan). If the borrower defaults, the lender must take his case to the courts and obtain a judgment. Only the courts have the power to sell the mortgaged property to obtain payment for the loan. Any surplus is returned to the borrower.

Real estate contracts usually are used where the borrower does not have sufficient down payment to qualify for a mortgage or the seller wants only a small down payment to qualify for installment reporting of capital gains income. To qualify for a mortgage, the borrower needs 25-35% down payment through regular commercial channels; whereas, under a contract he may borrow even if no down payment is made. However, the usual down payment with a contract is 10-25%. Under a contract the deed to the real property is held by the seller or placed in trust with a commercial firm such as a bank. Thus if the purchaser does not meet the conditions of the contract, such as being delinquent in payment, the legal processes for repossession of the property are much less restricting than for the mortgage holder. Frequently, the contract provides for giving of a deed and shifting to a mortgage when the borrower has paid 30-50% of the purchase price.

Personal property is financed through both secured and unsecured loans. Secured loans come under the Uniform Commercial Code in nearly all states. Chattel mortgages, conditional sales contracts, and other similar devices have been replaced. Even though they still may be used, their effect will be to create a "security interest." A *security interest* is an interest or legal right in a personal property or fixture to secure payment or performance of an obligation. To secure the loan or payment, the lender or seller requires the debtor or purchaser to file a "financing statement" or "security agreement" with the county recorder or secretary of state. A *financing statement* is a brief document describing the collateral furnished by the debtor. A *security agreement* is a loan agreement in which the collateral is described. To "perfect" a financing statement or security agreement, it must be filed. This notifies the public of the security interest the creditor has in the property of the debtor, thus protecting him from claims by third parties. Unperfected security interests are documents that are not filed. Perfected security interest established through the seller or a loan agency for the purchase of personal property is called a "purchase money security interest." For example, the seller or loan agency making money available to a farmer for the purchase of a tractor would have a purchase money security interest in the tractor purchased. Personal property given as collateral may be either tangible—such as consumer goods, equipment, and farm products—or intangible—such as promissory notes, bonds, corporate stock certificates, warehouse receipts, and security agreements.

This brief discussion of the Uniform Commercial Code is intended only to acquaint the reader with some of the provisions of Article 9 in the code which treats secured transactions. For a more complete discussion the reader is referred to Iowa Code, Ch. 554 (1971) or the respective code in his home state. The Iowa Code has ten articles that cover all phases of a commercial transaction involving personal property.

Unsecured personal property loans do not require collateral as security. A "signature or promissory note" is an example. This does not mean that the debtor has no financial obligation for the loan or that the creditor cannot bring about the sale of property through legal process for payment of the loan. However, secured loans take priority.

Farmers should provide a place in their file system for these credit instruments or documents as well as recording loan information in their credit accounts. Since many credit and sales agents retain the credit instruments in their files until the loan is paid, it is good practice to obtain duplicate copies. This gives the borrower a ready reference to the exact nature of his financial obligations. (A filing system is suggested in Chapter 13.)

INTEREST TABULATIONS

Interest is the chief cost for borrowing money and using credit, although other costs may be involved in negotiating loans. These include commission fees, recording fees, surveys, title certifications, insurance, and discounts.

When such charges are made, they become costs to the borrower and should be recorded in the credit account or elsewhere in the farm expense account.*

The method of charging interest makes a big difference in the cost of credit. There are many methods of tabulating interest charges; some of the more common are illustrated here. Also, the method of repaying the principal of the loan affects the true rate of interest paid. Thus both single-payment and multipayment situations will be illustrated for each method of tabulating interest.

Four factors are involved in tabulating interest costs: principal, time, rate, and payments. The *principal* is the sum of money loaned or borrowed. *Time* is the period within which the loan will mature. *Rate* is the percentage of the principal that is paid as interest and is usually expressed as a percentage per year. *Payments* refer to the number of installments in which the loan is repaid. These factors will be used in the discussion and examples that follow.

The true or real rate of interest is tabulated as follows:

$$\text{Rate} = \frac{\text{interest paid at the end of an interest bearing period}}{\left(\begin{array}{c}\text{outstanding balance at}\\\text{beginning of period}\end{array}\right) \text{x} \left(\begin{array}{c}\text{fraction of}\\\text{year borrowed}\end{array}\right)}$$

This formula will be used in illustrating simple interest tabulations in the following examples. All other interest tabulations are compared to this method as the true rate† For loans with regularly timed, equal principal payments the following equation is given as an approximation of the true rate of interest:

$$\text{True annual interest rate} = \frac{2 \text{ x} \left(\begin{array}{c}\text{number of regular payment}\\\text{periods in one year } (m)\end{array}\right) \text{ x} \left(\begin{array}{c}\text{interest payment}\\\text{or finance charges } (I)\end{array}\right)}{\left(\begin{array}{c}\text{loan amount received or}\\\text{unpaid balance of purchase } (B)\end{array}\right) \text{ x (number of payments } (n) + 1)}$$

$$= \frac{2mI}{B(n+1)}$$

This formula will be used in the examples that follow where the method of tabulating interest deviates from simple interest tabulations; i.e., discount and add-on interest loans.

1. *Simple interest* is paid on the outstanding amount of the money borrowed each time a principal payment is made.
   a. Single-payment loans: the total loan is retired in one single payment and the interest is paid when the loan becomes due.

   > *EXAMPLE 3.1*
   > Loan of $500 for 1 year at 8 % interest
   > Interest expense = $500 x .08 = $40

   > *EXAMPLE 3.2*
   > Loan of $500 for 6 months at 8% interest
   > Interest expense = ($500 x .08) x (6/12) = $20

---

*The 1974 edition of the Internal Revenue Service Farmer's Tax Guide, p. 21, states, "Loan expenses, such as legal fees,commissions, and other expenses paid in obtaining a farm loan, must be prorated and deducted over the terms of the loan."

†The Truth-in-Lending Act of Congress now requires firms selling on credit to inform the buyer of the true rate of interest.

b. Multipayment loans: the loan is retired through a series of principal payments, usually on a regular schedule. Interest is paid each time a principal payment is made and is tabulated on the outstanding (unpaid) amount of the loan after the last payment. This is known as an amortized loan.

EXAMPLE 3.3

Loan of $600 with principal payments of $50 made each month for 1 year; interest at 12% per year or 1% per month.

Period	Principal	Interest	Total payment	Unpaid balance
Beginning	. . .	. . .	. . .	$600.00
End of 1st month	$50.00	$600 x .01 = $6.00	$56.00	550.00
End of 2nd month	50.00	550 x .01 = 5.50	55.50	500.00
.				
.				
.				
End of 12th month	50.00	50 x .01 = 0.50	50.50	00.00

EXAMPLE 3.4

Loan of $10,000 with $1,000 annual principal payments; interest at 7%.

Period	Principal	Interest	Total payment	Unpaid balance
Beginning	. . .	. . .	. . .	$10,000
End of 1st year	$1,000	$10,000 x .07 = $700	$1,700	9,000
End of 2nd year	1,000	9,000 x .07 = 630	1,630	8,000
.				
.				
.				
End of 10th year	1,000	1.000 x .07 = 70	1,070	00

2. *Discount interest* is paid in advance on the original amount of the loan. The amount received is the amount of the loan less the interest charge.

a. Single-payment loans: the total amount of the loan (amount received plus interest) is retired in one single payment.

EXAMPLE 3.5

Loan of $600 for 6 months at 8% interest:

Interest paid = $600 x .08 x (6/12) = $24
Amount received = $600 − $24 = $576
Payment at the end of 6 months = $600
True rate of interest − ($24 ÷ $576) x 2 = .084 or 8.4%

b. Multipayment loans: payments normally are paid on a regular monthly basis over a time period of one year or less. But irregular payments for a period longer than one year and other than monthly intervals have been used.

EXAMPLE 3.6
    Loan of $900 at 7% interest with equal regular payments over a period of 10 months.

Interest paid = ($900 x .07) x (10/12) = $52.50
Amount received = $900.00 — $52.50 = $847.50
Monthly payments = $900 ÷ 10 = $90
Principal = $84.75
Interest = $5.25

True rate of interest = $\dfrac{2 \times 12 \times \$52.50}{\$847.50 \times (10 + 1)}$ = .135 or 13.5%

3. *Add-on interest* is added onto the principal amount of the loan and paid in equal installments with the principal payments.
   a. Single-payment loans: interest would be tabulated the same as for single-payment, simple interest loans. Hence this method is not applicable here.
   b. Multipayment loans: principal and interest payments are normally made on a regular monthly basis over a period of one year or less. However, periods of more than one year are fairly common.

EXAMPLE 3.7
    Loan of $1,200 at 9% interest for 12 months with even monthly payments.

Interest = $1,200 x .09 = $108
Monthly payment = ($1,200 + 108) ÷ 12 = $109

True rate of interest = $\dfrac{2 \times 12 \times \$108}{\$1,200 \times (12 + 1)}$ = .166 or 16.6%

EXAMPLE 3.8
    Loan of $4,000 at 7% for 36 months with even monthly payments.

Interest = $4,000 x .07 x 3 = $840
Monthly payments = ($4,000 + 840) ÷ 36 = $134.44
Principal = $111.11
Interest = $23.33

True rate of interest = $\dfrac{2 \times 12 \times \$840}{\$4,000 \times (36 + 1)}$ = .136 or 13.6%

REPAYMENT PROGRAMS FOR LONG-TERM LOANS

Repayment of loans made for one year or less are fairly straightforward and no additional discussion is needed here. Payments on loans with regular and even payments have already been illustrated under the discussion on interest in this chapter. However, there are a number of programs with irregularly timed equal payments, regularly timed unequal payments, and irregularly timed unequal payments, each tailored to a particular farmer's situation. The previous discussion on interest should be useful in developing these and understanding the manner of tabulating interest. This discussion on repayment programs will review common practices in developing long-term repayment programs.

Most long-term loans are retired on an amortized schedule of even payments. There are two common plans. The first, often referred to as the Springfield Plan, was illustrated in Example 2.4 under simple interest, multipayment loans. In this case the amount of principal payment was even (the same amount at each payment period), but the interest became smaller since it was charged on the unpaid principal balance. Thus the total principal and the interest payment become smaller with each successive payment. The other plan, often referred to as the Standard Plan, has an even total payment program. As the interest payment becomes smaller, the amount of principal payment becomes larger. The Standard Plan is illustrated in Example 3.9.

EXAMPLE 3.9

Assume a loan of $100,000 at 7% for 25 years with annual payments. From an interest table (Appendix Table A.4) it is seen that for a 25-year loan at 7% interest, the borrower will pay back $2.145263 principal and interest for each $1 borrowed. See Table A.3. Thus over the life of the loan the borrower will make principal and interest payments totaling $214,526. This is $8,581.05 per year over 25 years.

How much of the annual payment is principal and how much is interest? For the first year the tabulations are as follows:

$$\text{Interest} = \$100,000 \times .07 = \$7,000$$
$$\text{Principal} = \$8,581.05 - \$7,000 = \$1,581.05$$

For the second year:

$$\text{Interest} = (\$100,000 - \$1,581.05) \times .07 = \$6,889.33$$
$$\text{Principal} = \$8,581.05 - \$6,889.33 = \$1,691.72$$

The loan account for the first three years would appear as follows:

Period	Total payment	Principal	Interest	Unpaid balance
Beginning	. . .	. . .	. . .	$100,000.00
End of 1st year	$8,581.05	$1,581.05	$7,000.00	98,418.95
End of 2nd year	8,581.05	1,691.72	6,889.33	96,727.23
End of 3rd year	8,581.05	1,810.14	6,770.91	94,917.09

The differences between the Springfield Plan and Standard Plan are illustrated in Example 3.10.

EXAMPLE 3.10

Loan of $1,000 at 5% interest for 20 years with annual payments.

	Standard Plan				Springfield Plan			
	Annual payments			Unpaid	Annual payments			Unpaid
Year	Interest	Principal	Total	principal	Interest	Principal	Total	principal
Beg.	. . .	. . .	. . .	$1,000.00	. . .	. . .	. . .	$1,000.00
1st	$ 50.00	$ 30.34	$ 80.24	969.76	$ 50.00	$ 50.00	$ 100.00	950.00
2nd	48.49	31.75	80.24	938.01	47.50	50.00	97.50	900.00
3rd	46.90	33.34	80.24	904.67	45.00	50.00	95.00	850.00
	.	.	.	.	.	.	.	.
	.	.	.	.	.	.	.	.
19th	7.46	72.78	80.24	76.50	5.00	50.00	55.00	50.00
20th	3.82	76.50	80.24	00.00	2.50	50.00	52.50	00.00
	$604.80	$1,000.00	$1,604.80		$525.00	$1,000.00	$1,525.00	

The standard plan is usually better suited to beginning farmers whose money capital is limited.

The type of information illustrated in Examples 3.9 and 3.10 is readily available from commercial loan firms and should be provided to each borrower negotiating a long-term loan.

CREDIT FORMS

Repayment schedules of the various types of loans discussed in this chapter can be categorized as one of the following three types:

1. Single-payment loans, where the principal and interest will be paid in a single payment. These are usually for 3, 6, 9, or 12 months or some other combination that corresponds to a production or time period when some farm produce will be sold. Some single-payment loans may be for more than one year.
2. Multipayment loans, where the principal and interest will be paid by installments over a period of months and/or years. The distinguishing feature is not the duration of the loan but the division of payments. The payments may be of equal or unequal amounts and the time periods even or uneven.
3. Open-account loans, where items of supply are purchased but not paid for until the end of the month or some other billing period. These deferred payment purchases may even accumulate over a period of several months until some product is sold to furnish cash for the payment. Many farmers find it convenient to buy supplies under this arrangement and there may be some accounting advantages as well.

Each of these categories requires different accounting forms to record the needed information. For each it is desirable to have space for the (1) name of the loan agency, (2) purpose of the loan, (3) original amount and terms of the loan, (4) beginning date of the loan, (5) principal payments, and (6) interest payments. Noninterest loan costs such as insurance and processing can be recorded in the interest column of the loan account or in the expense account as discussed in Chapter 5.

The distinguishing features of the different loan accounts are in the provisions for recording loan payments. Loans with single payments need only one line to make these recordings. A single-payment loan form is illustrated in Form 3.1. Columns can be used to identify the items to be recorded. Loans with multipayments need a line for each

Form 3.1. Single-Payment Notes Payable for the Dunn Farm

Includes all single-payment loans, e.g., $500 for 6 months at 7% interest. Include in the interest column any noninterest loan charges. To allow more room to record loan detail, it may be convenient to use every second or third line.

	Date	Agency	Purpose	Original amount	Terms of payment	Payment		
						Interest	Principal	Date
1	Oct. 1, 71	Farmers State Bank	Purchase feeder pigs	$ 8000 00	7% for 1 yr	$ 560 00	$ 8000 00	Oct. 1, 72
2	Oct. 5, 72	P.C.A.	Purchase feeder steers	12000 00	7% for 1 yr			
3								
4								
5								
6								
7								
33								
34								
35	Column Totals			xxx	xxx	$560 00	$ 8000 00	xxx

40

Form 3.2. Multipayment Notes and Mortgages Payable for the Dunn Farm

	Description	Date paid	Total payment	Principal	Interest and carrying charges	Unpaid balance
						$ 55,000
1	Agency	Mar. 2, 60	$ ———	$ ———	$	55,000 00
2	Good Luck Insurance Co.	Feb. 28, 60	3240 00	490 00	2750 00	54,510 00
3		Mar. 4, 62	3240 00	514 00	2726 00	53,996 00
4	Purpose	Mar. 6, 63	3240 00	540 00	2700 00	53,456 00
5	Purchase 240-acre farm	Mar. 2, 64	3240 00	576 00	2673 00	52,889 00
6		Feb. 27, 64	3240 00	596 00	2644 00	52,293 00
7	Original amount $ 55,000	Mar. 5, 66	3240 00	625 00	2615 00	51,667 00
8	Date obtained March 5, 1960	Mar. 3, 67	3240 00	657 00	2583 00	51,010 00
9	Terms of payment:	Mar. 6, 68	3240 00	690 00	2550 00	50,320 00
10	5% standard	Mar. 2, 69	3240 00	724 00	2516 00	49,596 00
11	amortized loan	Mar. 5, 70	3240 00	760 00	2480 00	48,836 00
12	with annual	Mar. 3, 71	3240 00	798 00	2442 00	48,038 00
13	payments	Mar. 4, 72	3240 00	838 00	2402 00	47,200 00
14		73	3240 00	880 00	2360 00	46,320 00
15						

41

Form 3.3. Open Accounts Payable for the Dunn Farm

Includes open accounts at feed, fertilizer, and supply businesses where monthly statement is mailed.

		Date 1972	Added to account	Paid on account — Principal	Paid on account — Interest	Unpaid balance $ 574
1	Agency *Farmers Elevator Co.*	Jan. 1	$	$	$	$ 574 00
2		Jan. 6	216 00			790 00
3	Purpose	Jan. 23		574 00		216 00
4	*Purchase livestock feeds*	Feb. 6		216 00		0 00
5		Mar. 3	461 00			461 00
6		Apr. 23		461 00	4 61	0 00
7		Jun. 9	349 00			349 00
8		July 3	205 00			554 00
9	Terms of payment	July 8		349 00		205 00
10	*Cash in 30 days, 1% per*	Sept. 3	254 00	205 00	4 10	254 00
11	*month on accounts running*	Sept. 27		254 00		0 00
12	*over 30 days*	Nov. 15	310 00			310 00

Total interest $ 8 71

payment and provision for maintaining an unpaid (loan outstanding) balance. A multipayment loan form is illustrated in Form 3.2. Open-account loans need the additional provision for recording increases to the unpaid balance as well as payments. An open-account loan form is illustrated in Form 3.3.

The open-account loans record may not be needed under some accounting systems. Under the cash system of accounting where purchases are recorded only after payment has been made, it may be necessary only to maintain a business firm file of delayed payment purchases. If only one or two companies are involved, perhaps one unpaid accounts file is sufficient. However, under the accrual system where items are recorded when purchased whether paid for or not, an offsetting entry is needed in the credit section corresponding to the purchase entry in the expense section. However, whether the system is cash or accrual, for good credit management and control it is desirable to maintain an account for each business with whom an open account is maintained.

The credit account (all three types discussed) is also useful in other ways. It gives the amount of interest that has been paid as a business expense. This can be transferred directly to the income statement as a business expense and need not be recorded elsewhere in the accounts. The credit account is the basis for the liabilities side of the net worth statement. In addition, it provides information needed in developing a cash-flow statement—the basis of any good credit program or plan. Farmers need to plan the flow of cash into and out of their businesses to determine credit needs and investment opportunities. Cash cannot be used for business investment until fixed liability payments are made and family living needs are provided for. (The cash-flow budget will be discussed in Chapter 11.)

CASH AND ACCRUED INTEREST

The principal and interest time payments on loans do not usually correspond to the ending accounting period of the borrower. Thus the amount of cash interest payments does not usually correspond to the amount of interest that has accrued to the borrower. The amount of interest paid to service a loan is its cash cost; however, this may not correspond to the amount of interest the loan has generated. Interest accrues each day the loan is in existence, even on Sundays and holidays. Thus at any point in time there are unpaid interest balances accrued to the borrower. At the end of an accounting period when net worth is being established and net income is being tabulated, these accrued interest liabilities must be recognized and accounted for.

The difference between cash and accrued interest is shown in the following examples. The tabulation of unpaid interest liability is illustrated in Example 3.11 followed by accrued interest expense in Example 3.12.

EXAMPLE 3.11

Loan of $5,000 at 8% interest with annual principal payments of $1,000 and interest payments on the unpaid balance. The loan was taken out on April 1, 1970. The borrower is on a calendar year accounting period and the date is December 31, 1972. The loan account follows:

Form 3.4. Accrued Interest Tabulations for the Dunn Farm

Net Worth Statement

Column number (C):		(1)	(2)	(3)	(4)	(5)
Year	Loan identification	Unpaid balance	Annual interest rate (%÷100)	Months outstanding (since last payment)	Percent of year outstanding (C3÷12)	Accrued unpaid interest (C1xC2xC4)
1 1971	F.L.B. (Form 3.1)	$ 8000	.07	3	3/12	$ 140
2 1971	BPLC (Form 3.2)	48038	.05	10	10/12	2002
3 1971	Example 3.11	4000	.08	9	9/12	240
4						2382
5						
6						
11						
12 1972	PCA (Form 3.1)	12000	.07	3	3/12	210
13 1972	BfLC (Form 3.2)	47200	.05	10	10/12	1967
14 1972	Example 3.11	3000	.08	9	9/12	180
15 1972	Other	6300	varies			95
						2452

44

Form 3.4. (continued)

Income Statement

Column number (C):		(1)	(2)	(3)	(4)
			Accrued unpaid interest		
Year	Loan identification	Cash interest payment	End of this year, C5	End of last year, C5	Accrued interest expense (C1+C2−C3)
16	1972	JSB (Form 3.1)	$ 560	$ —	$ 140	$ 420
17	1972	Beq (Form 3.2)	0	210	—	210
18	1972	HXL (Form 3.2)	2402	1967	2002	2367
19	1972	Example 3.11	320	180	240	260
20	1972	Other	951	95	—	446
21			3633			3703
22						
23						

45

| | | Payment | | |
Date	Total	Principal	Interest	Total	
April 1, 1970	. . .	. . .	. . .	. . .	$5,000
April 1, 1971	$1,400	$1,000	$400	$1,400	4,000
April 1, 1972	1,320	1,000	320	1,320	3,000
April 1, 1973*	1,240	1,000	240	1,240	2,000

*Anticipated payment.

At the end of 1972 nine months have elapsed since the last principal and interest payment. Thus interest has accrued as an unpaid liability on the $3,000 principal balance outstanding after the last principal payment on April 1. The amount of accrued unpaid interest liability is tabulated as follows:

$$(\$3,000 \times .08) \times (9/12) = \$240 \times (3/4) = \$180$$

This $180 would appear as part of the total liabilities of the individual in his net worth statement at the end of 1972. The accrued interest expense for any one year is defined as the total interest generated on loans over the year whether· paid or not.

EXAMPLE 3.12

Assume the information given in Example 3.11. The interest expense for 1972 is made up of the interest accrued from January 1, 1972, to April 1, 1972, on the $4,000 and the interest accrued on the $3,000 from April 1, 1972, to December 31, 1972.

$$(\$4,000 \times .08) \times (3/12) = 320 \times (1/4) = \$\ 80$$
$$(\$3,000 \times .08) \times (9/12) = 240 \times (3/4) = \underline{\ 180}$$
$$\text{Total annual accrued interest} = \$260$$

Whereas the accrued interest expense for 1972 is $260, the cash interest payment was $320.

Form 3.4 was developed for the purpose of tabulating accrued interest for both the net worth statement and the income statement.

4

INVENTORY ACCOUNTS AND NET WORTH STATEMENT

NET WORTH STATEMENT

The net worth statement is a summary of the financial assets of the business and the claims upon them. Assets not claimed by firms outside the business are the owner's residual claim or equity. The form of the net worth statement is illustrated as follows:

Net worth statement

Assets			Liabilities
$			$
$			$
$			$
$			$$$
$$$$		Owner's equity	$
			$$$$

The owner's equity is often referred to as the net worth. This gives rise to the basic accounting formula:

$$\text{Assets} = \text{liabilities} + \text{net worth (owner's equity)}$$

The net worth statement is also referred to as a balance sheet. This arises from the basic accounting formula — the assets are either claimed by an outside business interest or by the business owners. This balance or equality is always maintained because the owner's equity is a residual claim (Fig. 4.1).

The assets of the business are the productive resources that have economic value. To have economic value, resources must be scarce and capable of rendering future services to the business. As such these items can be bought and/or sold on the market.

47

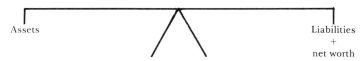

Fig. 4.1. Basic accounting formula.

Examples of farm assets are land, fences, wells, buildings, machinery, equipment, livestock, crops, fuel, fertilizer, receivables, and cash. To list the assets is to describe the farm in terms of its financial worth. These are summarized on the left-hand side of the net worth statement or balance sheet. The categories into which they usually are grouped are current assets, working assets, and fixed assets.

Current assets are the supplies, raw materials, and merchandise being held for sale. Examples are repair parts, feed, fertilizer, fuel, feeder livestock, and crops being held for sale. These are commonly known as the inventory in double-entry accounting, which will be discussed later. For convenience, cash and receivables could also go into this class. Looking at it another way, current assets are cash or near cash and either are consumed in the production process or are being held for sale.

Working assets support the production process with their services but are not used up in one accounting period and are not directly and legally attached to land. Through the giving of services they become less valuable and thus are an expense to the business. These may be looked upon as prepaid business expenses to be charged as their value changes through time. Examples of working assets are machinery, equipment, tools, and breeding livestock. Many of these assets can be found on the business depreciation schedule, which will be discussed later in this section.

Fixed assets are land or attached to land. Examples are farmland, building sites, wells, terraces, fences, and buildings attached to land. Land normally does not depreciate as it renders services as do buildings and fences. Some land improvements depreciate, whereas others do not.

A typical farm net worth statement is shown in Table 4.1. This same farm will be illustrated in other places in the text. Compare the asset and liability classes and consider the types of items included in each.

Many textbooks do not separate working assets from fixed assets and thus have only two classes. Fixed assets are defined to include working assets; however, in this text they are separated to facilitate the analysis that follows. Also, from a legal and financial basis they are different. On the tax records, real property is separated from personal property. The financing of land is considerably different from financing machinery or livestock and is often handled by separate agencies.

The separation between classes of assets for many items is not clear and may even be arbitrary; sometimes the separation may be more convention than real. There is something to be said for consistency and comparability. Unless the same definitions are used from one accounting period to the next, the financial changes cannot be accurately measured. If records of investment, income, and production are to be compared among farms, they must have the same basis for comparison. Two useful tests can be applied when classifying assets. The first is the convertibility test and the second is the use or service test.

The *convertibility test* relates to the ease with which the asset can be converted into cash at its market value. Many items have well-established market outlets and the price is broadly known. Grain and feeder livestock are examples. Other items such as farm

TABLE 4.1. Net worth statement for the Dunn farm, 1971

Assets			Liabilities		
Current:			Short-term:		
Cash	$	3,100	Feed bills	$	574
Life insurance (cash value)		500	Cattle note		8,000
Savings bond		518	Interest on farm mortgage		2,382
Feed		16,519			$ 10,956
Nonfeed crops and supplies		2,120			
Feeder hogs		9,130	Intermediate:		
Feeder cattle		10,337	Auto loan (farm share)	$	1,037
		$ 42,225			$ 1,037
Working:			Long-term:		
Machinery and equipment	$	8,514	Silo	$	4,000
Gilts			Farm mortgage		48,038
Sows		3,038			$ 52,038
Portable buildings		548			
		$ 12,100			
			Total Liabilities		$ 64,031
Fixed:					
Farm land at $420/A		$ 96,000	Net Worth		$103,146
Buildings and improvements		16,852			
		$112,852			
Total Assets Owned		$167,177			
Landlord Assets Rented		$ 62,000			
Total Assets Managed		$229,177			

machinery are regularly traded, but the markets are not so well established and prices are less well known or understood. Thus it may take considerably more time and effort to find the highest price for a used machine. The land market, which includes improvements, is the least well established and there may not be a separate market for some improvements. Many farms are on the market for a year or more waiting for a buyer who is seeking those peculiar resources and is able to come to terms with the seller.

The *use test* has to do with the nature of the services the assets render to the business. Most current assets are completely used up in the production process within the accounting period, or they are sold. Most working assets provide services over several accounting periods and thus become less valuable. Fixed assets such as soil provide services indefinitely and may even become more valuable in the process.

Most farm lenders require a good accounting of current assets within a farm business because most of these assets will be used or sold within the accounting period, which normally is one year. The returns from the use and sale of the current assets indicate the potential cash flow within the business during the coming accounting period. There are certain demands on this cash income such as living expenses, retirement of short-term liabilities, interest payments on all liabilities, and operating expenses of the business. Lenders like to see the short-term liabilities cleaned up annually, and they look toward the current assets to see whether there will be enough cash flow to meet these payments.

The working or intermediate-term assets are also of concern to lenders. First, these assets are not as readily available for sale in the accounting period. Some of the return or the cost is recovered through use and depreciation. Machinery must be replaced at some point in time; there will also be some turnover of breeding stock. Although there may be some sales out of working assets such as breeding stock or used machinery, normally these will be replaced. For instance, if some breeding animals are sold, some young breeding stock usually will be transferred from current assets to working assets. Thus sale of breeding stock may be balanced by a decrease in the amount of livestock that can be sold from those listed in current assets. Likewise, as a depreciated machine is replaced, it normally involves a considerably higher expenditure than is listed on the depreciation schedule or the financial statement. This machine not only has depreciated but many times it is replaced by a machine that costs somewhat more than the machine being replaced. This is due to several factors such as inflation of machinery costs as well as upgrading of machinery either in size, power, or features.

It should be emphasized again that the assets are the physical resources of the business (farm) and do not relate to ownership. The legal right to use an asset makes it an asset, even though legal title has not been passed. Rented assets, however, should be separated from owned assets since the farm operator does not have ownership control over them.

Liabilities represent the financial claims upon the owned assets. Normally these are the unpaid debts of the business. Examples are accounts payable, notes payable, mortgages, accrued unpaid interest, and wages payable. Liabilities are classified into categories that correspond to their respective assets, i.e., short-term, intermediate-term and long-term.

Short-term liabilities correspond to the current assets and include notes, unpaid accounts, interest payable, and principal payments falling due within the following accounting period.

Intermediate-term liabilities correspond to the working assets and include notes and security interest agreements with principal payments running for more than one accounting period — usually more than one year and less than ten.

Long-term liabilities are mortgages and contracts on real property. These usually run 10-30 years from their initiation.

The residual claim belongs to the owners and is referred to as the owner's equity or net worth. It is shown on the right-hand side of the balance sheet with the outside liabilities. If the assets are looked at from a corporate sense as an individual, the owner's equity is a form of liability in that he too has a claim upon the assets. In this text liabilities will refer to the debts of the owner, and owner's equity or net worth will be the owner's residual share.

For the time spent the net worth statement provides a greater amount of financial information than any other account a farmer may keep. The net worth statement refers to a point in time and does not record the dynamic changes between net worth statements. It is a static account much like a photo that stops everything in motion and records it as it is at the moment. It forms a benchmark against which all financial progress is measured, and it is the foundation of most credit transactions.

FARM INVENTORY

The term inventory can be used as a noun or a verb. As a noun it refers to property. In nonagricultural businesses the property is restricted to goods held for sale, goods in the process of production, and raw materials. Durable goods such as depreciable property and land normally are not included. However, in farm accounting

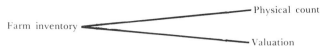

Fig. 4.2. Taking inventory.

all personal and real property commonly is included in inventory. In legal terminology inventory refers to a complete list of all assets, both current and noncurrent.

As a verb, inventory refers to a process of counting and valuing resources. This two-part verb usage generally will be used in this chapter. The working definition to be used is: inventory is an accounting of all property owned or controlled by the farmer operator including a listing of all liabilities. The process of listing and valuing, however, will be concerned mostly with crops, livestock, and productive supplies. Depreciable property is already listed and valued in the depreciation schedule discussed in Chapter 2, and land normally is carried over at the value it held in the opening inventory.

Taking the inventory includes two processes, a physical count and a valuation (Fig. 4.2). Making the physical count is a practical matter and does not justify lengthy discussion. The only equipment required is a notebook and pencil and a measuring device such as a rule and scale. Each piece of property is listed and adequately described; however, several points should be kept in mind:

1. It is important to measure quantities of items in units that are commonly used and meaningful and that aid in establishing values, e.g., bushels for small grains, hundredweight for market livestock, gallons for fuel, etc.
2. Like items should be grouped according to their contribution to the business or their place in it. For example, all livestock should be grouped under one classification and further divided by type, age, and/or purpose. Crops should be grouped and divided by type and differentiating properties. Farm machinery should be grouped and divided by purpose or use. It may be desirable to separate powered from nonpowered machinery or the livestock equipment from the crop equipment, etc. Whatever divisions are made should be meaningful and for a purpose. This will be dictated by the farm analyses the operator wishes to make. The groupings should fall into the categories established for the net worth statement.

Converting of measured quantities into meaningful units is sometimes troublesome. It may be necessary in the inventory process to measure the quantity of grain on hand in cubic feet, count the number of bales of hay in the barn or measure the cubic feet, measure the depth of corn in the silo, etc. Tables and conversion formulas are available that facilitate this process. Even though the measurement under this method is not exact, it is usually accurate enough for practical purposes. Conversion tables are available in many farm account books. (See Appendix Table A.2.)

Some items in the inventory are included in the farmer's depreciation schedule discussed in Chapter 2. For these items the physical count will be one of adjustment for items added or disposed of during the year.

The inventory of rented properties should also be included. Rented resources are not greatly different from assets purchased with borrowed money and should be counted. They are part of the farmer's assets in the sense that he has full or partial control over their use. A value for these may or may not be given depending on the use and analyses being made of the inventory account. Note in the example net worth statement (Table 4.1) that the rented assets are shown on the asset side but not on the liability side and are not included as part of the owned assets.

When valuing farm inventories, the farmer should keep in mind several important ideas:

1. The major purpose of the inventory is to obtain a true picture of net worth. For this to reflect the current net worth position of the farmer, current market prices must be used. However, this may conflict with other purposes for taking the inventory.
2. The inventory reflects noncash income flows. A crop produced may not be sold but carried over for later sale and thus cause the crop inventory to increase, or a crop on hand in the previous inventory may be sold and not replaced. Changes in value of durable inputs such as machinery and buildings are treated as business expenses. Thus the inventory should accurately reflect net farm income concerning receipts and expenses as well as changes made possible through new investment and debt retirement.
3. If properly analyzed, the inventory is useful for management purposes. The physical count and values applied should not distort this analysis and render it meaningless.
4. The inventory needs to be legal. One of the major reasons for keeping records is to meet the requirements of the IRS. Thus the methods applied in valuing the inventory must fall within the specifications of the law. A broad interpretation of this requirement would include lease arrangements, loan agreements, farm estates, etc., where some legal or moral implication is involved.

Common methods used in valuing farm properties are as follows:

1. Market cost
2. Net market price
3. Farm production cost
4. Cost minus depreciation
5. Cost minus depletion
6. Capitalization

The cost referred to in the methods above may include more than just the price at the point of purchase. The cost at the farm may include transportation, site preparation, installation, etc. These must be added to the cost of the item, particularly where the service life of the asset is more than one accounting period.

The *market cost value* of an item refers to the price at which it was purchased on the market. This method works particularly well for items that have been recently purchased and will be used up within a relatively short period of time. Examples might include feed concentrates, fuel, small tools, fertilizer, seed, and feeder livestock. For feeder livestock it would be necessary to adjust the cost for changes in value (production costs) after purchase.

The *net market price* is the value of an inventory item at the marketplace less transportation and marketing charges. In other words, it is the money the farmer could bring home from the sale of a product. This method works particularly well for items that are held primarily for sale or for which a market price is well established. Examples include market livestock and farm-produced crops. This method may be useful in particular situations for valuing land, machinery, breeding livestock, and other items not being held for sale. Farms are often appraised on the basis of comparative market sales.

"Cost or market, whichever is lower" is a method closely related to the two

discussed above. The cost of producing an item, or its original purchase price, is compared to the current market price and the lower of the two amounts is chosen. The major advantage of this method is conservatism. However, its user may make mistakes when prices are increasing as well as decreasing. The method is probably too conservative when prices are increasing and not conservative enough when they are falling, but these are problems of either of the above methods also.

The *farm production cost* method relates to the cost of producing an asset on the farm. The production cost should not include profit, interest on investment, or the opportunity cost of the operator's labor and management. Generally this method is most useful when valuing assets that have been produced on the farm and will be used in other farm production activities. The best example is breeding livestock that have been produced on the farm. The term "unit-livestock-price method" has been used by the IRS. Under this method the livestock are classified according to age (i.e., calves, yearlings, two-year-olds, mature animals) and each is given a value as estimated by its production cost. The IRS requires that all raised livestock be valued by this method if it is adopted and that, once established, the values cannot be changed without IRS consent. Example values might be the following:

Calves (3-9 months)	$ 60
Yearling heifers (10-15 months)	85
Two-year-old heifers (16 months to calving)	120
Mature cows	150

When applied to purchased livestock, the value used is the purchase price adjusted for the age change in value for raised animals.

Silage and standing crops present particular valuation problems. The production cost method may be a possibility, but often the costs of production are vague and illusive. It may be possible to value silage in terms of equivalents of other crops for which a price is established. For example, 1 ton of silage and 4 bushels of corn are roughly equivalent to $\frac{1}{3}$ ton of hay. Another method may be to value the corn silage on the basis of the estimated corn grain equivalent adjusted for the difference in harvesting costs.

Standing crops may be valued in terms of their estimated yield multiplied by the expected price adjusted for harvesting costs and uncertainty of harvest. A practical approach to value will solve most of these difficult problems.

The *cost-minus-depreciation* method pertains to investment properties that provide services to the business for a period longer than one year and become less valuable in the process. These items take the form of prepaid expenses in that the initial purchase price is not all used up in one year but is allocated over the life of the asset. Assets valued by this method are buildings, machinery, fences, and purchased breeding livestock. The cost of these items are recovered from future revenues spread over the life of the asset. For example a tractor may last 10-15 years. Part of the value of the purchase price remains each year but becomes smaller through time. The change in value from one accounting period to the next becomes a business expense. The value of the asset is its value at the beginning of the accounting period minus the change in value from one accounting period to the next. This process of tabulating a schedule of annual value changes is called depreciation and is discussed in Chapter 2.

Cost minus depletion refers to valuing a stock asset such as a gravel pit. The value of the asset changes in accordance with the amount of resource displaced or removed during the accounting period. Timber stands, mineral deposits, oil wells, and organic deposits are examples of these resources.

Capitalization as a valuation method has reference to time problems. Land provides income over a presumably infinite period. An endowment provides income over a finite period of time. An unpaid sale may not be collectable for a year, if at all. Some enterprises produce only after several years of sequential investments, thus their cost of production is an accumulated one. The cost of time enters into these value problems. The right to collect $100 one year from now is not worth $100 now; waiting is an expense. Moreover, investing money one way prohibits it from being invested in another. The cost of investing in the first way is the return that could have been received by investing in the second. These problems will be investigated more fully in a later section in this chapter.

Internal Revenue Service Restrictions

1. Items subject to depreciation must be valued by the cost-minus-depreciation method except for purchased breeding livestock.
2. Once a valuation method has been selected and applied, special permission is required to change to another method.
3. If the "cost or market, whichever is lower" method is used, it must be applied separately to each item.
4. If the net-market-price method is used it must be applied to the entire inventory except for breeding livestock, which may be valued by the unit-livestock-price method or placed on the depreciation schedule.
5. Livestock purchased for resale may not be valued by the unit-livestock-price method, which is restricted to raised livestock and livestock purchased for breeding purposes. If the cost method of valuing purchased feeder livestock is used, the value selected should include the farm costs incurred after the date of purchase.
6. Livestock purchased for breeding purposes may be added to the depreciation schedule or valued by one of the more direct methods — cost, market, or unit-livestock-price method. Persons using the cash method for determining taxable income should place purchased breeding livestock on their depreciation schedules since the depreciation allowable, whether claimed or not, must be deducted from the purchase price when determining capital gains income.
7. Under the unit-livestock-price method the following rules apply:
   a. Once the unit prices have been selected they must be maintained at the same level for raised livestock until permission from the IRS is obtained to change them.
   b. Immature purchased breeding livestock are valued at their purchase price adjusted for farm costs of production incurred after purchase. However, no adjustments are made if the livestock are purchased during the last six months of the taxable year. The farm production costs are those reflected in the unit prices selected. Mature animals are maintained at their purchase price.

Capitalization

Discounting is the process of converting a future income into a present value. It is a time consideration and reflects the cost of waiting. For example, suppose a farmer sells $100 worth of livestock to a neighbor and agrees to wait one year for his payment. The problem comes in determining what this future income is worth in the current inventory. One way to look at this is to consider the farmer's cost of waiting one year to receive his payment. If he had the $100 now he could put it to use productively on the farm or off, and if he is a good manager, it should grow to be worth over $100 within the year, perhaps to $110. Alternatively, he may need to borrow $100 more for which he must pay $6-$7 interest. Thus there is a *real cost* of waiting. Another way to look at it is to ask, How much would the farmer need to deposit in a savings account or invest in his business to have it grow to be worth $100 at the end of a year? The answer is in terms of the expected rate of interest return. (See Appendix Table A.3.)

The formula given for calculating the present value of a future income amount is as follows:

$$V = R/(1 + r)^t$$

where V = present value, R = expected future income, r = estimated rate of return (interest rate), and t = waiting time in years. Applying this to the problem situation and assuming the rate of return is 5%, the present value is

$$V = \$100/(1 + .05)^1 = \$100/1.05 = \$95.24$$

If the agreed payment is $100 in each of the next three years, the present value is

$$V = \frac{\$100}{(1 + .05)^1} + \frac{\$100}{(1 + .05)^2} + \frac{\$100}{(1 + .05)^3}$$

$$= (\$100/1.0500) + (\$100/1.1025) + (\$100/1.1576)$$

$$= \$95.24 + \$90.70 + \$86.39 = \$272.33$$

Farmland poses the problem of an infinite series of future incomes. When a farmer buys a farm, he purchases the right to receive the flow of income from the land in perpetuity; and while the farmer surely will not live that long, his hope is to sell the farm for as much or more than he paid when he wishes. Mathematically the above formula converts to the simple formula used in almost all farmland appraisals:

$$V = R/r$$

where R now represents the average expected net return from the land resource. This is usually interpreted to mean the residual income from the land after all expenses have been paid and resources rewarded. This method is illustrated in Table 4.2 using an actual farm appraisal.

The interest rate becomes all important when applying the capitalization method. This is particularly true with respect to land. The three most common philosophies are as follows:

1. Select a rate representative of the rate of return that could be received in alternative investments; i.e. municipal bonds, government securities, or preferred stocks.
2. Select a rate that converts the anticipated land income into a value which approaches market sale prices. This is the method used in the example appraisal.
3. Select a rate that corresponds to the farm mortgage rate of interest.

TABLE 4.2. Value of a central Iowa farm calculated by the capitalization method, using the landlord approach and assuming typical management

Income source	Acres	Yield	Rental rate	Rental share	Price	Owner's income
Corn	100	105	1/2	5,250 bu	$1.30	$ 6,825
Beans	40	36	1/2	720 bu	3.25	2,340
Oats	10	65	1/2	325 bu	.75	244
Hay	15		cash		25.00	375
Timber	30.8				0	0
Lots and house	3		cash		60/mo	720
Roads	6				0	0
Owner's Estimated Gross Income						$ 10,504

Estimated expense under typical operation:

Taxes (assessed value = $29,520; mill levy = 85)	$ 2,460
Insurance on buildings (rate $.40/$100 value on $14,905)	60
Maintenance on buildings (rate = 3%, value = $14,905)	450
Seed	325
Fertilizer	875
Combining corn and beans (½ of $8/A)	560
Herbicide	160
Miscellaneous (fences, management, tile, improvements)	400
Owner's Estimated Gross Expense	$ 5,290
Owner's Estimated Net Earnings ($10,504 − $5,290)	$ 5,214
Earnings Value:* $V = R/r = \$5,214/.038$	$137,210
Earnings Value per Acre for the 204.8 Acres	$ 670

*The r value of 3.8% reflects the current relationship between land sales and the anticipated net land earnings in the community.

The first two alternatives are the most reasonable. The mortgage rate of interest is a cost for borrowing money and not a return on an investment. Which of the first two approaches to choose is a matter of choice based upon the purpose of the appraisal. The first approach gives a value based upon the opportunity cost of investing in the farm. The second gives a value more nearly equaling current market prices.

Where the future income flow is constant and forthcoming over a period of several years, a combination of the last two formulas may be useful.

$$V = \frac{R}{r} - \frac{R/r}{(1 + r)^t}$$

Consider an annuity of $100 per year over the next 10 years at 5% interest:

$$V = \frac{\$100}{.05} - \frac{\$100/.05}{(1 + .05)^{10}} = \$2,000 - \frac{\$2,000}{1.6290} = \$2,000 - \$1,228 = \$772$$

Income from buildings, machines, and breeding livestock may be examples where this method might be useful. Income annuities through life insurance companies are calculated using these same considerations.

Closely related to discounting procedures discussed above are the *compounding of costs*. In this case a present value of a past cost is calculated. The formula is

$$C = c(1 + r)^t$$

where C = present value of past cost, c = past cost, r = rate of return, and t = number of years since the cost occurred. For illustration consider an orchard tree planted four

TABLE 4.3. Compounding of costs

Number of years	Past cost	Formula	Present value
4	$10.00	$C = 10 (1 + .05)$	$12.15
3	2.00	$C = 2 \ (1 + .05)$	2.32
2	2.00	$C = 2 \ (1 + .05)$	2.21
1	2.00	$C = 2 \ (1 + .05)$	2.10
	$16.00		$18.78

years ago and a rate of return of 5%. The present accumulated real cost of planting the tree and nurturing it to the present is not the cash outlay of $16 but $18.78. If the money had not been spent for the tree, it could have been invested elsewhere at a return. See Table 4.3.

In relation to capitalization of future income payments, it may be important to consider uncertainties of collection. A percentage discount is allowed by most retail business firms for collection fees and nonpayment of outstanding credit.

Preferred Valuation Method

Table 4.4 lists the inventory valuation methods previously discussed and the items commonly found in the farm inventory and indicates which methods may be most useful when placing a value on each item. It will be noted that for many items more than one method of valuation can be applied. The methods selected will not be the same for all farm businesses. The following items should be considered when selecting the method or methods to use:

1. The values established will be entered in the net worth statement. If the objective is for this statement to reflect current net worth, market value methods should be selected.
2. The values established will be used in evaluating the efficiency of the business if an analysis is made. For example, inventory values for crops may be used in tabulating the value of crops produced per acre and value of feed fed, which will be used in figuring returns per $100 feed fed.
3. Requirements of the IRS must be met for property used in tax reporting. Most farmers do not keep separate inventories for business analysis and tax reporting.
4. The principles of ''conservatism'' and ''consistency'' should be considered.
5. The principles of ''the going concern'' and ''stable monetary unit'' should be considered, especially for land. (See Chapter 1 for a discussion of the principles.)

Changes in land values due to price fluctuations probably should not be made in the inventory between the opening and closing inventory dates. Capital gains on landholding are realized only when the land is sold. This gain may be realized only once in the lifetime of a farmer. Up until this time any gains or losses are only on paper because they have not been converted to cash. It probably is good practice to value land and similar assets conservatively and to hold these values nearly constant over several years of inventory. Changes should be made between the closing inventory of one period and the opening inventory of the next and should be reflected as capital gains or losses, not farm income.

TABLE 4.4. Useful methods to consider when valuing various kinds of farm assets in the inventory process

Assets	Purchased cost	Net market price	Lower of cost or market price	Farm production cost	Cost minus depreciation	Replacement cost minus depreciation	Discounting of future returns	Compounding of past cost
Cash		X						
Savings bonds		X					X	
Raised feed		X		X				
Purchased feed	X							
Nonfeed crops		X	X					
Feeder livestock		X	X					
Breeding livestock		X	X	X				
Receivable sales							X	
Receivable loans							X	
Powered machinery					X			
Nonpowered machinery					X			
Equipment					X			
Small tools	X					X		
Portable buildings					X			
Fixed buildings					X	X		
Fences					X	X		
Terraces					X			
Orchard	X				X		X	X
Land		X			X		X	

Fig. 4.3. Value of a brood cow.

Difficult Valuation Problems

Most valuation problems can be dealt with in a reasonable fashion by considering the valuation methods discussed and the purpose of the net worth statement. Four illustrations indicate how these might be handled:

1. Land improvements may be of a depreciable or permanent nature. If the improvement is land leveling, it may add permanently to the value of the land. If it wears out such as a terrace might, it is a depreciable asset. Items to consider when valuing the improvement are (a) land preparation costs such as surveying, grading, bulldozing, etc.; (b) the cost of idling the land while the improvement is being made, i.e., taxes, interest, weed control, etc.; (c) government sharing of costs;* and (d) cost of the actual improvement, i.e., reseeding, tiling, terracing, etc.

2. Rented assets should be valued in order to tabulate the total assets managed. They are part of the farm much the same as assets purchased on credit. The simplest method is to value them separately and show them only on the asset side of the net worth statement. Less exactness is required in the valuation process.

3. Livestock present peculiar problems since they appreciate as well as depreciate. The value of a typical commercial brood cow may vary as shown in Figure 4.3. The unit-livestock-price method allows this variation, as does the net market price method. However, regular depreciation methods do not, and there is no easy way to make the adjustment. The taxpayer filing his taxes by the cash basis is particularly affected where purchased immature breeding stock are involved. He must value his livestock for tax purposes at their cost less depreciation, which may not have any resemblance to market value. Thus it may be necessary for some taxpayers to have two valuation schedules, one for taxes and one for their own management use.

4. The effect of changes in the general price level is important to consider in relation to the net worth statement. Price level changes affect the asset values and the equity of the owners. Thus it is important to consider the changes in the valuation process. However, the prices for some assets continually change. If these enter into the valuation process, they cause the net worth to increase and decrease for reasons other than productivity of the business. It is generally considered a good practice to hold constant between the beginning and ending inventories the value of land, breeding livestock, and other assets not being held for sale or on the depreciation scheddule. Changes in capital values can be shown as capital gains between the ending of one inventory and the beginning of the next.

---

*Usually government financial assistance must be added to income in the year received, and the total cost capitalized. The manner in which soil and water expenditures are handled on income tax returns also may be a factor.

Form 4.1.   Land Inventory for the Dunn Farm

> Prices for land can be shown by land quality of individual tracts or for the total farm. Values may be arrived at from comparative sales or capitalization or may be based upon general values published by the agricultural experiment station or cooperative extension service. Land values need not be retabulated annually. Capital gains should not be added to the annual income statement.

Owned Land (excluding buildings)

	Acres	Description	Use	Value per acre	Total value
1		Farm H		$	$
2	192	Rotated cropland	Row crops and hay	430	82,560
3	30	Improved pasture	Continuous	250	7,500
4	18	Farmsteads, fences, roads			5,940
5					
6					
7					
8					
9					
10					
11					
12					
13	240	Totals (show average value per acre)		$ 420	$ 96,000

		Value per acre	Total value
14	Capital gain (loss) over value shown in previous year's inventory	$ none	$ none

Rented Land (including buildings rented)

	Acres	Description	Use	Value per acre	Total value
15		Farm R - crop share			
16	77	Rotated cropland	Row crops	400	30,800
17	3	Waste		—	
18					
19		Farm C - cash rent		400	29,600
20	74	Rotated cropland		}	1,600
21	3	Waste			
22	3	Farmstead			
23		Totals (show average value per acre)		$	$ 62,000

Form 4.2. Feed Crop Inventory for the Dunn Farm

	Description	Unit	Quantity	Price	19 71 value		Quantity	Price	19 72 value	
	Supplies include fuel, fertilizer, twine, etc.									
1	Grains Corn	bu	13,100	$ 1.04	$ 13,180	00	15,520	$ 1.07	$ 16,296	00
2	Oats	bu	2,200	.65	1,430	00	2,500	.65	1,625	00
3										
4										
5	Forages Hay	ton	60	18.00	1,080	00	56	18.00	1,008	00
6										
7										
8										
9	Concentrates Beef suppl.	cwt	82	4.50	370	00	64	4.75	316	00
13	Swine supplement	cwt	85	5.40	459	00	72	5.50	396	00
14	Column Totals	xxx	xxx	xxx	$ 16,519	00	xxx	xxx	$ 19,641	00

Form 4.3. Swine Inventory for the Dunn Farm

Entries are for the end of each year. See tax regulations for valuing livestock and be consistent from year to year. Livestock on depreciation will be recorded in this section and also be shown on the depreciation schedule.

Separate livestock on depreciation from other livestock and value them at their undepreciated values. Do not double-count values.

	Description	19 71				19 72			
		No.	Total weight	Price	value	No.	Total weight	Price	value
1	Old Sows	54	18,900	$ 53/aa	$2862 00	55	19,250	$ 53/aa	$2805 00
2	Gilts					10	2500	60/aa	600 00
3	Boars	2	700	98/aa	176 00	2	600	100/aa	200 00
4							200	75/aa	75 00
7	Breeding Stock Total	56	19,600	xxx	$3039 00	68	22,550	xxx	$3680 00
8	Market Hogs	32	7,680	14.00	1075 00	26	6,240	15.00	936 00
9		381	52,630	15.30	8055 00	253	27,830	16.50	4584 00
10									
14	Feeding Stock Total	413	60,310	xxx	$9130 00	279	34,070	xxx	$5520 00
15	Total Farm	469	79,910	xxx	$12168 00	347	56,620	xxx	$9200 00

Forms for recording the inventory should provide space for:

1. Describing the property
2. Recording the quantity
3. Showing the price selected
4. Listing the total value

These forms may only have spaces for recording one year's inventory or they may have spaces for several years. Multiyear forms save transferring information from one year to the next and facilitate business analysis comparisons. Forms 4.1, 4.2, and 4.3 are examples of inventory forms. [Depreciation forms are shown in Chapter 2 and liability (credit) forms are shown in Chapter 3.]

NET WORTH STATEMENT ANALYSIS

The general concepts of the net worth statement and balance sheet have already been introduced. Interpretation of the statement can probably be done best through an illustration. Table 4.5 shows the net worth statement for the Dunn farm for 1972. It shows net worth statements for both the beginning and the ending of the year. Note that the ending of one year is the beginning of the next. Next look at the various categories of assets and liabilities and note the changes that have taken place over the year. Feeder cattle have increased, whereas feeder hogs have decreased. However, there are $600 worth of new gilts on hand ready for production the next year. The increase in cattle loans has caused the total liabilities to increase by about $4,000.

Compare asset and liability balances to see if there are sufficient assets to cover the corresponding liabilities. Most respected loan agents recommend that borrowers not borrow on fixed assets to cover current expenses. Sufficient current assets should be on hand to cover the liabilities that will fall due before new production takes place and also to provide for family living over the same interval. The asset to liability ratios are useful in this regard. The *net capital ratio* (NCR) is the most useful calculation and compares dollars of total farm assets per dollar of total farm liabilities. In Table 4.5 the 2.6 beginning net capital ratio was calculated as follows:

$$\text{NCR} = \text{total assets/total liabilities} = \$167,177/\$64,031 = 2.6$$

Similar ratios can be calculated for the current and working assets. The working asset ratio includes the current assets and short-term liabilities in addition to working assets and intermediate liabilities.

What might be termed a safe capital ratio has not been established. This relates somewhat to credit. A farmer may be able to borrow up to 100% of his equity to buy feeder cattle or feed, perhaps 80% to buy machinery, and 65% to buy land. Assuming a 30% equity safe, this would give a net capital ratio of 100/70 = 1.4. Certainly a 1.0 ratio is unsafe and 2.0 would be safe in most situations; but what might be considered a safe ratio between these extremes varies with farmers, farms, and the nature of prices. Good farm managers can use more credit than poor ones, some farming activities are more stable than others (e.g., yearling steer feeding and cash-grain), and stable prices would

TABLE 4.5. Net worth statement for the Dunn farm, 1972

Assets	Beginning	Ending	Liabilities	Beginning	Ending
Current:			Short-term:		
Cash	$ 3,100	$ 2,800	Feed bills (3.3)	$ 574	$ 476
Life insurance (cash			Cattle note (3.1)	8,000	12,000
value)	500	550	Interest, accrued		
Savings bond	518	544	unpaid (3.4)	2,382	2,452
Feed (4.2)*	16,519	19,641		$ 10,956	$ 14,928
Nonfeed crops and sup-					
plies	2,120	2,492	Intermediate:		
Feeder hogs (4.3)	9,130	5,520	Auto loan (farm share) $	1,037	$ 518
Feeder cattle	10,337	14,080	Truck		2,300
	$ 42,225	$ 45,627	Tractor		3,000
			Grain bin		1,000
Working:				$ 1,037	$ 6,818
Machinery and equip-					
ment	$ 8,514	$ 16,080	Long-term:		
Gilts (4.3)		600	Silo	$ 4,000	$ 3,000
Sows (4.3)	3,038	3,080	Farm mortgage (3.2)	48,038	47,200
Portable buildings	548	470		$ 52,038	$ 50,200
	$ 12,100	$ 20,230			
			Total Liabilities	$ 64,031	$ 71,946
Fixed:			Net Worth	$103,146	$106,963
Farmland at $420/A (4.1)	$ 96,000	$ 96,000	Change in Net Worth	$3,817	
Buildings and im-					
provements.	16,852	17,053			
	$112,852	$113,053			
Total Assets Owned	$167,177	$178,909		Beginning	Ending
Landlord Assets Rented (4.1)	62,000	62,000	Capital ratios:		
Total Assets Managed	$229,177	$240,909	Net capital ratio	2.6	2.5
Change in Total Assets			Working capital ratio	4.5	3.1
Managed	$11,732		Current capital ratio	3.8	3.1

*Numbers in parentheses indicate form numbers where figures are developed.

allow for a lower ratio than fluctuating prices. Thus what constitutes a safe ratio must be determined individually. The banker also may have some ideas of his own.

The ratio of current assets to short-term liabilities is usually of special concern for lenders with operating or short-term loans. Generally lenders would like a ratio of 2 to 1, i.e., $2 of current assets to $1 of short-term liabilities. Such a ratio, or even higher, is usually identified with a relatively safe loan. The Dunn farm in Table 4.5 has a much more favorable ratio than $2 of current assets to $1 of short-term liabilities.

Lenders would also like to see a working capital ratio of $2 of assets to $1 of liability. This ratio will vary somewhat with the type of enterprise. For instance, a farm with a specialized beef-cow or dairy enterprise may have a high percentage of livestock listed in working assets and may have a heavier loan on these than is characteristic of some other assets. Therefore, the ratio may be different in this type of business from one that has market hogs or cattle feeding. There is no set rule on ratios; therefore this becomes a guideline that the lender and the borrower must establish. The change in ratio of assets to liabilities over time may be just as significant as the ratio at any given time. This information over a period of several years will indicate the financial progress of the business over time. This will provide some guidelines for judging whether or not the business is making satisfactory financial progress.

The balance of assets to liabilities and the size of the net worth are useful in some of the more personal aspects of the business. For example, the amount and kind of crop

TABLE 4.6. Relationship of capital investment to farm size, farm type, and income efficiency

Item	Farm size*			Farm type†			Efficiency†		
	160 acres	320 acres	640 acres	Grain farms	Hog farms	Dairy farms	High profit	Low profit	Dunn farm
	(%)			(%)				(%)	
Livestock and feed	26	30	32	12	19	22	27	21	26
Machinery and equipment	9	7	6	7	9	10	8	9	9
Land and improvements	65	63	62	82	73	68	65	70	65
	100	100	100	100	100	100	100	100	100

Source: 1971 Iowa Farm Business Summaries, Iowa Coop. Ext. Serv., Ames, 1972.
*Northwest Iowa.
†All farms.

insurance needed may be determined by the total assets managed in relation to the liabilities. Also, the amount of life insurance needed for family security is directly related to the stability of the business and the size of the net worth.

The figures for total assets managed are also of interest. These relate to the size of the business and are reached by adding the value of rented assets to the total assets owned. Changes representing growth can be measured by comparing the amounts for the beginning and ending of the year.

Changes in net worth also can be measured. This should be of particular interest to the farm family. This change is made possible only if farm income is more than necessary to provide for family living if the farm is the only source of income. Thus, there is a direct relationship between farm income and farm growth and ownership. These comparisons are not particularly significant in any one year due to the nature of agricultural production and prices. However, a comparison over time should have real meaning.

The investment balance of a business may be useful to consider. Some investment relationships for Iowa compared to the farm illustrated in Table 4.5 are shown in Table 4.6. There is an economic optimum investment balance for every farm. What that balance should be may be elusive, and it will change over time. A comparison with farms of similar size and type in the same geographic location may be revealing. This comparison may not answer many questions but it could raise some worth investigating.

The optimum investment balance relates to the size, type, and location of the farm as well as to the age and financial security of the operator. Some of these differences are shown in Table 4.6. Smaller farms tend to have a greater percentage of investment in livestock. Young farmers also tend to have greater investment in livestock than older farmers as well as less debt. Capital turnover is important to the beginning farmer.

5

RECEIPT AND EXPENSE ACCOUNTS AND INCOME STATEMENT

SINCE IT IS REQUIRED BY LAW that farmers pay taxes on their net income, all farmers keep a semblance of a receipt and expense account. In some cases this function is turned over to professional accountants or calculated on electronic data processing machines. Regardless of who performs the function or how it is done, the account must reflect the income position of the farm business. The Internal Revenue Service (IRS) has stated it "does not prescribe any particular kind of record book; such books should, however, be a systematic record that will accurately and clearly reflect the income, deductions, credits, and other matters required to be shown on the income tax returns. The keeping of a few casual notes or records, which will enable you to do little more than approximate your income, is not adequate." The case for complete and accurate records has already been stated. The purpose of this section is to describe procedures and methods of accounting in sufficient detail to allow the determination of farm income and the fulfillment of the objectives stated above.

The net worth statement reflects the assets and liabilities of the business, while the income statement summarizes the receipts and expenses. In a sense the receipt accounts can be looked upon as the production accounts since they reflect the sale of the items produced on the farm. The expense accounts reflect the service aspects of the business — the cost for labor, machinery, feed, fertilizer, utilities, insurance, and similar items.

ACCOUNTING SYSTEMS

Most farmers use the single-entry system of accounting; most other businesses use the double-entry system. Both methods are acceptable and if ac-

66

curately kept will give the same income statement. The principle of the single-entry system is that receipt and expense transactions are recorded in only one place. The double-entry system rests upon the principle that there is both a source and destination to every transaction and thus at least two entries must be made. Emphasis here will be upon single-entry accounting. There are two general systems of recording data in the single-entry account, the cash and accrual methods.

Under the *cash method* entries are made in the accounts only after the cash or money has exchanged hands. A purchase is recorded when it is paid for and not when the transaction takes place, if at a different time; a receipt is recorded when the money is received, not when the sale was transacted, if at a different time. For example, fuel purchased in December 1971 but not paid for until January 1972 will be recorded as a 1972 expense. Inventory changes reflecting crop and livestock increases or decreases are not included because they have not been converted to cash. Depreciation is counted as an expense. In the long run all these deferred purchases and sales and inventory increases and decreases will show up as cash, and income over time will be accurately described.

The *accrual method* follows the approach that income should be accurately described each year. Thus transactions are recorded when binding whether money changes hands or not; changes in inventory are included. The fuel purchase described above is recorded as an expense for 1971 when delivery was received, not in 1972 when the payment was made. If a beginning inventory is converted into cash, this shows up as a decreased inventory (expense) as well as a cash receipt. A crop that is raised but not sold is likewise shown as an increase in inventory value and recorded as a receipt.

Which of these two methods a farmer selects is a matter of choice; but if he is interested in analyzing his farm business for management purposes, he cannot tolerate the transfers of production and expenses and fluctuations of income inherent in the cash method. However, there are advantages and flexibilities in the cash method compared to the accrual method when calculating taxable income. Thus it is logical for some farmers to allow for both in their accounting system. (The merits of each method for taxation purposes will be discussed in Chapter 10.) It is logical that cash income can be derived more easily from the accrual method than accrued income from the cash method; hence the accrual method will be emphasized in this chapter.

INCOME STATEMENT FORMS

To demonstrate the differences between the cash and accrual methods of keeping the single-entry income account, we will look at the *cash-flow income account*. Most everyone is familiar with the monthly bank statement issued to each depositor. (See Form 5.1.) This is a cash-flow account used to inform depositors of their cash balance at the close of the monthly accounting period and shows all the transactions that have affected the beginning cash balance to arrive at the ending cash balance. Every check (bank draft) written on the account is shown and every deposit, from whatever source, is shown. The depositor can use this statement to compare with his own records, such as his checkbook stubs and business receipt and expense accounts, to determine if he (or the bank) has made any recording errors or omissions.

In addition to being useful for reporting balances and checking errors, the statement gives considerable information about the business. The size of the business can be estimated from the size of the individual transactions. The volume of business

Form 5.1. Example Monthly Bank Account Statement

STATEMENT OF YOUR ACCOUNT WITH

Union Story Trust and Savings Bank

"YOUR FRIENDLY MAIN STREET BANK"

OFFICE AT NORTH GRAND SHOPPING CENTER AMES, IOWA

OFFICE AT GILBERT, IOWA

A FULL SERVICE BANK

ACCOUNT NO.

02/13/73
DATE

10

ERRORS SHOULD BE REPORTED IN TEN DAYS

HANDLE ALL YOUR FINANCIAL TRANSACTIONS AT OUR FULL SERVICE BANK

USE REVERSE SIDE FOR RECONCILING YOUR ACCOUNT

NEW CAR LOANS AT 9.3 PER CENT SIMPLE

01716   LAST STMT.

BALANCE LAST STATEMENT	NO.	CHECKS AND DEBITS AMOUNT	NO.	DEPOSITS AMOUNT	SERVICE CHARGE	BALANCE THIS STATEMENT
3,178.10	40	2,327.95	2	1,230.70	.00	2,080.85

DATE	CHECKS AND OTHER DEBITS		DEPOSITS	BALANCE
01/17	51.68			3,126.42
01/18	8.01	10.00		
	31.70	57.24		3,019.47
01/22	5.00	15.00		
	35.46	50.00		
	1,000.00			1,914.01
01/23	20.00	100.00		1,794.01
01/24	4.25	15.00	7.98	1,782.74
01/25	10.88			1,771.86
01/26	12.53	20.00		1,739.33
01/29	3.00	22.96		1,713.37
01/31			1,222.72	2,936.09
02/01	1.95	13.56		
	193.30			2,727.28
02/02	30.90			2,696.38
02/05	1.23	3.34		
	10.00	40.60		2,641.21
02/06	9.00	9.00		
	15.00	220.00		2,388.21
02/07	20.00			2,368.21
02/08	3.64	20.00		
	48.10	159.33		2,137.14
02/12	9.07	12.32		2,115.75
02/13	15.00	19.90		2,080.85
40				

TX - STATE TAX
SC - SERVICE CHARGE
DM - DEBIT MEMO
CM - CREDIT MEMO
LS - LISTED CHECKS
OD - OVERDRAFT
RT - RETURN CHECK
DC- DEPOSIT CORRECTION
EC - ERROR CORRECTION

PLEASE EXAMINE AT ONCE. SEE REVERSE SIDE FOR AN EASY RECONCILIATION FORM

transactions can be seen by counting the number of monthly transactions. If this volume becomes very large, it may require the service of an accountant to record them. The minimum monthly balance may give some indication of business stability, and the difference between the beginning and ending balance may tell something about business growth. Form 5.1 shows 40 withdrawals and 2 deposits. The minimum balance was $1,713.37, the largest withdrawal was $1,000, the largest deposit was $1,222.72, and the cash balance was decreased by $1,097.25. It would be very risky to draw many conclusions from the study of one month's statement; a study of 12 statements would be more reliable and revealing. The months when deposits and withdrawals were the largest would stand out, reflecting seasonal production and sales. A study of these statements over a number of years would be even more revealing and might provide sufficient evidence to a banker to justify a loan:

As useful as a bank statement might be, it is not too helpful in analyzing the productive activities of the farm business. It was designed to report the financial condition of the bank account and not as a substitute for farm financial records. The farmer should duplicate every bank transaction someplace in his financial accounts. If a withdrawal was made to purchase a capital asset, it should show up in the farmer's capital purchase account; if it was to pay off a loan (make a principal payment), it should be recorded in the farmer's credit accounts; or if it was to pay for operating expenses such as labor or fertilizer, it should be recorded in the business operating expense account. Similarly, if a deposit was from the sale of livestock, it should be recorded in the livestock sales account; but if it were from a bank production loan, the deposit should be added to the credit account.

An account could be developed that would keep the same financial data as the bank statement but would identify the sources of the deposits and the destinations of the withdrawals. Such an account is shown in Form 5.2 and is called a cash journal. Data are recorded chronologically, and a running cash balance is maintained. This account is very similar to the one many people keep with their checkbook and adds considerably to the bare figures shown on the bank statement. With this it is possible to associate receipts and expenditures with business activities. Whether the farm is for livestock or crops can now be determined, and whether the deposits were from sales of produce or capital or from new loans is distinguishable. Expenditures can be associated with the family or farm business, and business expenses can be categorized by activity or purpose.

But there are still difficulties in analyzing the activities of the business with the cash journal account. If one wishes to determine the total sales of livestock or the total money paid to hired labor or for feed, it is necessary to sort through the many transactions and make a separate list by category of the several item analyses wanted. This may become a tedious process. Often a check (bank draft) is written to cover the purchase of several items that may be unrelated, such as gasoline and a candy bar or bolts and fertilizer. The cash journal may not give this detail. A further complication is that often checks are written to obtain pocket cash, and the cash is then used to purchase various business or personal items and may go unrecorded.

Assuming that it is possible to itemize and accumulate the many cash transactions over one accounting period (usually one calendar year), a cash-flow income statement could be summarized as illustrated in Table 5.1. Note that cash expenditures may be for farm operation, livestock and feed purchases, fixed expenses, capital purchases, and loan payments including business account payments. Receipts include sales of farm products, transfer payments from government, refunds and dividends and other

Form 5.2. Cash Journal for Recording Deposits and Withdrawals from a Bank Account

CHECK NO.	DATE	CHECK ISSUED TO	AMOUNT OF CHECK		✓	DATE OF DEP.	AMOUNT OF DEPOSIT		BALANCE 3,126 42	
341	1/15	Jones Implement	51	68	✓				3,074	74
342	1/15	Internal Revenue Service	1000	00	✓				2,074	74
343	1/16	Mel's Service – Cash	10	00	✓				2,064	74
344	1/18	Farm-Home Utility	31	70	✓				2,033	04
345	1/18	County Treasurer - auto license	57	24	✓				1,975	80
346	1/18	Faraway Grocery	35	46	✓				1,940	34
347	1/18	Mastercharge account	50	00	✓				1,890	34
348	1/21	Mel's Service – cash	20	00	✓				1,870	34
✓	1/21	Transfer - Fredah's book	100	00	✓				1,770	34
348	1/22	Clark's Hardware	4	25	✓				1,766	09
✓	✓	Life insurance refund	—	—	–	1/22	7	98	1,774	07
349	1/22	Pete's Electric - 5.00 cash	15	00	✓				1,759	09
350	1/25	J. C. Penney - farm	12	53	✓				1,746	54
351	1/25	J. C. Penney - family	10	88	✓				1,735	66
352	1/25	Standard Oil Co.	3	00	✓				1,732	66
✓	✓	Livestock Auction Co.	—	—	–	1/25	1,222	72	2,955	38
353	1/26	Smith Lumber	22	96	✓				2,932	42
354	1/26	Oscar Langy, DVM	20	00	✓				2,912	42
355	1/30	Clark's Hardware	1	95	✓				2,910	47
356	1/31	Coop Elevator - acct.	193	30	✓				2,717	17
356	2/1	Standard Oil Co.	3	34	✓				2,713	83

miscellaneous farm income, sales of capital assets, and new cash loans. Since this is the farm cash-flow account, transfers to the operator and his family are not shown directly. The net cash-flow amount reflects cash that has been available for family living, nonfarm investments, and changes in the cash bank balance and cash asset account. This is important information to compute. The accuracy of the farm cash accounts can be checked by comparing the farm net cash-flow income with the totals from these other records.

TABLE 5.1. Cash-flow income statement for the Dunn farm, 1972

Cash expenses		Cash receipts	
Farm operation:		Farm production:	
Hired labor	$ 4,293	Livestock sales	$53,939
Livestock expenses	1,182	Crop sales	3,012
Crop expenses	4,543	Livestock product sales	0
Fuel and oil	1,163		
Machinery repair	1,464	Government payments:	
Building repair	354	Feed grain program	2,436
Machine hire	1,682	ASCS	0
Farm utilities	232		
Miscellaneous	247	Other:	
		Refunds and dividends	370
Feed purchases	9,635	Custom work and wages	0
Livestock purchases	12,880		
		Sale of capital assets	0
Fixed expenses:			
Taxes	1,584	New farm loans:	
Insurance	334	Capital purchase	3,000
Interest on loans	3,633	Farm operation	10,000
Cash rent	2,000	Total	$72,757
Capital purchases:			
Cash paid down	4,400	Cash-flow summary:	
		Cash receipts	$72,757
Principal payments:		Cash expenditures	59,983
Farm loans	10,357	Net	$12,774
Total	$59,983		

To make the cash journal more functional and to reduce the time and increase the accuracy of making category summaries, a multicolumn journal account could be used. (A multicolumn account is illustrated in Form 5.4 and will be discussed in more detail where the various record-keeping forms are presented in this chapter.) A column could be added for each item for which a separate figure is wanted. Thus livestock expenses, crop expenses, labor hired, feed purchases, etc., could be recorded in separate columns and summarized at the end of each month and/or year. If a joint farm-family bank account is maintained, it would be desirable to have a column for recording family cash transfers. (The details of a cash journal account are discussed elsewhere in this chapter and are illustrated in Form 5.9.)

Net farm income—cash method. Over a long period the cash-flow account and summary cash-flow income statement will reflect the productive income from the business. However, for any one accounting period (year) it may be very misleading and not reflect the value of production or its cost.

First, consider new loan deposits and principal loan payments. The money received from a new loan may be used to finance farm production (such as to buy feed, pay labor, and make insurance payments) and income from farm sales may be used to make principal payments, but loan receipts themselves are not productive. Likewise, loan payments are not business expenses. To include as expenses both the loan payment and the purchase of the goods for which the loan was contracted would be double-counting. New loans or loan payments do not affect the business equity position since the liability and asset accounts are affected by equal amounts.

Second, the purchase of a farm capital asset with a life longer than one year is not totally a business expense since the asset still has recoverable value at the end of the

TABLE 5.2.   Cash income statement for the Dunn farm, 1972

Cash expenses		Cash receipts	
Farm operation:		Farm production:	
Hired labor	$ 4,293	Livestock sales	$53,939
Livestock expense	1,182	Crop sales	3,012
Crop expense	4,543		
Fuels and lubricants	1,163	Government payments	2,436
Machinery repair	1,464	Value of livestock products	
Building repair	354	used in the home	320
Machine hire	1,682	Other	370
Utilities (farm)	232	Total	$60,077
Miscellaneous	247		
		Farm cash income summary:	
Feed purchases	9,635	Cash receipts	$60,077
Livestock purchases	12,880	Cash expenses	49,837
		Net	$10,240
Fixed expenses:			
Taxes	1,584		
Insurance	334		
Interest	3,633		
Cash rent	2,000		
Depreciation:			
Machinery	3,734		
Improvements	877		
Total	$49,837		

accounting period. The change in value that takes place from the date of purchase (or the beginning of the accounting period) to the end of the accounting period is the production expense. (This was discussed in Chapter 1 as depreciation.) Similarly, the sale of a capital asset does not constitute a farm product sale. A capital asset sale merely converts the market value the asset had into cash. This is a more liquid form of wealth, but the owner is no wealthier. Also added is the value of farm products used in the home. This is an offsetting entry to balance the expenses that have been charged for their production. A profit should not be reflected in this item.

An income statement with these modifications is shown in Table 5.2. This statement is now very near the one many farmers use to report their taxable income from farming by the cash method. The major difference is that the IRS requires that the purchase of feeder livestock not be shown as an expense until the year in which they are sold. In some cases the purchase and sale take place within the same accounting year, but more often not. In Table 5.2 if the livestock sold as part of the $53,939 receipts had been purchased for $14,000, the expenses would be $50,957 ($49,837 − $12,880 + $14,000) and the cash net taxable income would be $9,120 ($60,077 − $50,957). (Other considerations when tabulating taxable income will be discussed in Chapter 10.)

Net farm income—accrual method. Over time the cash farm income statement reports productive income quite accurately. Any increases in inventory caused by annual production not being sold in any one year will be offset over a period of years by sales of products out of inventory that were produced in a prior accounting year. The only unreported production value will be the long-run inventory increases of salable product assets such as livestock and crops. In any one year there may be purchases not paid for and hence unrecorded as expenses (such as feed and fuel and vice versa), but over time these too will balance out. But for any one accounting period the cash income statement may grossly misstate the true productive income—the net value increase of the

TABLE 5.3. Accrual income statement for the Dunn farm, 1972

Accrual expenses		Accrual receipts	
Operating expenses:		Farm production:	
Hired labor	$ 4,293	Livestock sales	$53,939
Livestock	1,182	Crop sales	3,012
Crops	4,543		
Fuel and lubricants	1,163	Government payments	2,436
Machinery repairs	1,464		
Building repairs	354	Other	370
Machine hire	1,682		
Utilities (farm)	232	Inventory increases:	
Miscellaneous	247	Livestock	775
		Crops and supplies	3,494
Feed purchases	9,371		
Livestock purchases	12,880	Value of livestock products	
		used in the home	320
Fixed expenses:		Total	$64,346
Taxes	1,584		
Insurance	334	Farm accrual income summary:	
Interest	3,703	Receipts	$64,346
Rent	2,000	Expenses	49,643
		Net	$14,703
Depreciation:			
Machinery	3,734		
Improvements	877		
Inventory decreases:			
Livestock	0		
Crops	0		
Total	$49,643		

business. Inventory amounts may vary greatly from one accounting period to the next, and some rather large purchases can be made at the end of one accounting period that will not be billed and paid for until the next. For net income to reflect the true net value increase of the business, these noncash business transfers must be accounted for.

The net income statement tabulated by the accrual method includes cash and noncash receipts and expenses and accounts for inventory changes of productive enterprises. An income statement tabulated by the accrual method is shown in Table 5.3. Note how it differs from the cash income statement shown in Table 5.2:

1. Inventories were included, and increases exceeded decreases by $4,269.
2. Feed purchases were less than feed account payments by $264 ($9,635 − $9,371).
3. Interest expenses were $70 more ($3,703 − $3,6333). (For an accounting of interest tabulations by the cash and accrual methods, see Form 3.4.)

It is assumed that all other purchases and sales not otherwise identified were for cash or the deferred payment purchases or sales were paid for or receipted within the accounting period.

The accrual income statement is sometimes referred to as the production income statement because it more nearly reflects the total value of production and the costs of obtaining that production. Each statement is accurate for the year for which it is tabulated; thus one year can be compared with another to measure any chages in production, production requirements, prices, and income. This is the only true income statement of those discussed and thus is the most useful for business analysis and decision making.

There is no set format for an income statement, and many businesses arrange their data to make it useful for management purposes. One format that finds wide use among farmers relates to the cash flows in the business. Operating expenses are separated from fixed expenses, and cash expenses are separated from noncash expenses. Receipts are handled similarly. The format for such an income statement is shown in Table 5.4. Note that the terms *debits* and *credits* are used. For our purposes when referring to the income statement, debit will be used synonomously with expense and credit with receipt. Debits and credits refer to both cash and noncash transactions and noncash increases and decreases in product inventory accounts. Debit and credit terms grew out of the system of double-entry accounting and will be defined more completely in a later section of this chapter. Expenses and receipts have been used sometimes to refer to cash transactions only, but our use of these terms will be broader and will refer to both cash and noncash items. The concern under the accrual method is to account for changes in the worth of the business and its operator. Thus, under an accrual account those items

TABLE 5.4. Entries in an accrual income statement

Debits	Credits
Cash Expenses	*Cash Income*
Operating:	Livestock sales
Livestock expense	Livestock product sales
Crop expense including fertilizer	Crop sales
Machinery and equipment repair	Machine hire
Improvements, repair, maintenance	Misc. receipts (govt.. refunds. dividends, etc.)
Fuel, oil, grease	Machinery and improvements, profit or loss from
Hired labor	sale
Farm utilities	
Misc. farm (dues, literature, office, etc.)	*Noncash Income*
	Inventory increase: Livestock, crops, supplies
Feed: Grain and hay purchased	Farm products used in the home
Livestock purchased:	*Total Business Credits*
Poultry	Cash income *plus* noncash income
Hogs	
Cattle	*Total Cash-Flow Farm Income*
Other	Cash income *plus* undepreciated value of machinery, equipment and improvements sold,
	new loans, and outstanding balance of accounts
Fixed expenses:	
Taxes	INCOME MEASURES
Insurance	
Interest	*Gross Profit*
Rent	Total business credits *minus* inventory decreases:
	Crops. supplies livestock: livestock purchases; and feed purchases
Noncash Expenses	
Depreciation:	
Machinery and equipment	*Net Farm Income*
Buildings and improvements	Total business credits *minus* total business debits
Inventory decrease: Livestock, crops, supplies	*Net Cash-Flow Income*
	Total cash farm income *minus* total cash expenditures
Total Business Debits	
Cash expenses *plus* noncash expenses	
Total Cash Expenditures	
Cash expenses *plus* cash paid down on capital purchases, principal payments on loans, and beginning balance of accounts	

labeled "cash" may not always involve immediate cash transfers such as with an open business account.

Cash expenses are for farm operation, the purchase of feed and livestock, and payment of fixed expenses. Fixed expenses are separated out because they are constant and do not relate to changes in the productivity of the business as do operating expenses. They would remain nearly the same even if the farm were not operated.

Noncash items include depreciation and inventory changes of crops, livestock, and production supplies. Depreciation accounts for the change in value from using machinery, buildings, equipment, and other depreciable items in the business. Since the sum of depreciation charged over an asset's life may not be equal to its total change in value (purchase price minus sale value), it may be necessary to adjust the depreciation amount by the difference between the undepreciated balance and the sales value. This adjustment is shown under cash income as "machinery and improvements, profit or loss from sale."

The total of cash and noncash expenses equals *total business debits*. The total of cash and noncash receipts equals *total business credits*. Total business credits minus total business debits equal *accrual net farm income*.

Net income measures:

1. The return to unpaid (operator and family) labor
2. The return to management
3. The return to unpaid (net worth) capital

All other items have been paid a return for their use in the business. Market prices have been charged for operating expenses and fixed cash expenses; depreciation has been charged for machinery and buildings, etc.; and interest has been charged for liability capital used in the business. Only the above three items have not been paid for their contribution to the business.

Net farm income reflects the monetary value the family had (or has) available for living, debt retirement, increasing equity in the business, or other investment or savings. Part of this income may be reflected in the inventory of livestock or crops, which is a form of investment within the business. Where there was an inventory decrease of livestock and crops, the cash-flow income would be greater than the accrual net income because more of the inventories were sold than replaced. The family should realize in this case that if the business investment is to be maintained, this monetary value difference should be invested back into the business. Accrual net farm income is a reliable guide for measuring the financial progress of the business. It is from this income that investment and equity in the business grows.

Accrual net farm income as defined above produces a figure similar to the one used for income tax reporting under the accrual method. However, because of capital gains treatment of certain sales and other differences, the reportable income for income tax purposes may be somewhat different from the net income as reported in the farm business analysis. Since the accounting procedures and regulations for income tax purposes do not follow completely acceptable accounting procedures, one should determine the net farm income independent of tax responsibilities.

Net cash-flow income was discussed earlier and illustrated in Table 5.1. It is possible to tabulate either this or the cash net income (Table 5.2) from the accrual net income statement. Only a few adjustments are necessary. These can be seen by comparing Tables 5.1 and 5.3 for the net cash-flow income. The first apparent difference is that noncash expenses are not included in net cash-flow income. Thus we can begin with a summary of the cash expenses and cash income. Cash paid out for capital purchases and loan principal payments must be added to the cash expenses and adjustment made for open-account differences and accrual and cash interest payments. It will be recalled that accrual expenses are recorded even if the item was not immediately paid for but was added to the business account. To obtain the cash payment from the accrual expense account, it is necessary to adjust for the difference between the beginning and ending business account balances. Note that in Table 5.1 feed purchases were $264 larger than in Table 5.3. This indicates that $264 worth of feed ($9,635 − $9,371) was purchased in a previous accounting period (1971) and not paid for until the current accounting period (1972). Next note that in Table 5.1 interest payments were $3,633, whereas in Table 5.3 an expense of $3,703 was shown. The $70 difference is due to the fact that more interest had accrued than had been paid for—the interest liability had increased by $70. Money received from new farm loans and the undepreciated value of capital assets sold must be added to the cash income on the credit side of the accrual income statement. Since the profit or loss from sale of capital assets has already been recorded, only the undepreciated balance need be added. The undepreciated balance plus the difference between it and sale value is equal to the total dollars received from sale.

Gross income and *gross profit* are often calculated from the income statement to measure gross volume of business and value of all items produced on the farm. Gross income is calculated by subtracting from total business credits any inventory decreases of crops, supplies, and livestock. It should be realized that when calculating total business credits these same items were considered for any inventory increases. Gross income is a good measure of farm productivity and business volume especially for grain farms. But when feeder livestock is a major enterprise or farm raised grains are purchased for livestock feed its use is not reliable for comparing different farms or even the same farm over time. Gross profit adjusts for these two items.

Gross profit is tabulated by subtracting from gross income the purchase costs of feeder livestock and livestock feeds. The removal of the purchase cost of livestock from gross income leaves the value of livestock increase for the farm being measured (livestock sales − livestock purchases + (closing inventory value − beginning inventory value) = value of livestock increase). For example, consider two farmers who had the same livestock sales, but one produced his own feeders while the other purchased his. If this adjustment were not made, they would each have the same gross profit from livestock production even though the purchased feeders were raised on some other farm. The subtraction of livestock feeds purchased from total business credits is not so clear. Feed purchases are removed to adjust for difference among farm feeders and to give consistency to the income measure.

Consider two farmers who had identical farms and operations except that one farmer fed home-grown grain to his livestock while the other sold his grain and purchased a complete feed. If these two farmers are to have similar gross profit figures (as they should), an adjustment needs to be made for the difference in obtaining feed for their livestock because the second farmer had larger grain sales and hence a larger gross income. If the livestock feed purchases are subtracted, the gross profit figures for the two farmers should be very similar. The farmer who had the larger grain sales should also have the larger feed purchases. In summary, gross profit is calculated by subtracting from total business credits inventory decreases of crops, supplies, and livestock; livestock purchases; and feed purchases.

Gross profit also can be defined as the total value of crops produced plus the increase in value of livestock over their total feeds fed value *plus* miscellaneous income. (Chapter 7 discusses how gross profit can be tabulated from production records.)

Table 5.5 illustrates these income statement concepts as applied to the Dunn farm in the net worth statement of Table 4.5. It should be noted that this statement is divided into three columns for the operator, landlord, and total farm. The "total farm" column reflects the situation that would exist if the farm were completely owner operated. Thus the crop-share rent payment would be available for sale, and there would be no cash rent transfers. The leasing arrangements illustrated are those typical for a midwestern crop-share lease.

The differences between the several income tabulations discussed in this section are illustrated in Table 5.6. The format deviates from those used previously in order to emphasize the differences of the income statements. Notice the way deferred payment purchases are handled as illustrated in feed purchases; how interest expense tabulations differ between accrual and cash accounting; how capital asset purchases and sales are accounted for; and what inventory information is used. It should be observed that the ending value of machinery inventory is equal to the beginning inventory *plus* purchases *minus* the undepreciated balance of sales *minus* depreciation.

RECEIPT AND EXPENSE ACCOUNT FORMS

Forms used to record receipt and expense transactions are of three general types:

1. Cash journals
2. Multicategory accounts
3. Single-category accounts

A *journal account* is a single-account listing of all financial transactions as they occur in the business. Receipt transactions are separated from the expenditure transactions by page or column; i.e., receipts could be recorded in a separate account book or in different sections of the same book from the expenses or they could appear in the same account book with the dollar amounts shown in separate columns. An example of this kind of an account is shown in Form 5.3. Farm and family transactions can be combined or separated. In the example, family living expenditures are shown as transfers from the farm but are not detailed. Some businesses record only cash transactions, while others record all business transactions, whether cash was transferred or not. In the example the transactions that did not reduce the bank balance are placed in parentheses.

The principal merit of the journal account is in keeping track of the total cash flows of the business. However, it is not much better than a check-stub record in this respect. It does provide the basic format for all other receipt and expense accounts. Its major drawbacks are that it does not differentiate between kinds of receipts and expenses, it does not make it easy to keep track of physical amounts or per unit prices, and it does not help in recording deferred payment transactions that do not have an immediate cash flow. Thus this kind of record does not facilitate analysis of the production and financial

TABLE 5.5. Income statement for the Dunn farm, 1972

Item	Operator	Landlord	Total farm
INCOME MEASURES			
Cash Income			
Hog sales (5.5)	$30,809	. . .	$30,809
Cattle sales	23,130	. . .	23,130
Feed crop sales	202	. . .	202
Nonfeed crop sales	2,810	. . .	2,810
Value of crop-share rent	. . .	$2,426	2,426
Cash rent	. . .	2,000	. . .
Government payments	2,436	337	2,773
Miscellaneous receipts	370	0	370
	$59,757	$4,763	$62,520
Noncash Income			
Inventory increase:			
Crops, feed, and supplies (4.2)	$ 3,494	. . .	$ 3,494
Livestock (4.3)	775	. . .	775
Value of farm products used in home	320	. . .	320
	$ 4,589	$ 00	$ 4,589
Total Business Credits			
Cash income	$59,757	$4,763	$62,520
Plus noncash income	4,589	. . .	4,589
	$64,346	$4,763	$67,109
EXPENSE MEASURES			
Cash Expenses			
Operating:			
Hired labor (5.8)	$ 4,293	. . .	$ 4,293
Livestock expense	1,182	. . .	1,182
Crop expense (5.7)	1,976	$ 125	2,101
Fertilizer, lime (5.7)	2,567	400	2,967
Fuel, lubricants	1,163	. . .	1,163
Machine and equipment repair	1,464	. . .	1,464
Buildings and improvements repair	354	60	414
Machine hire	1,682	. . .	1,682
Utilities (farm share)	232	. . .	232
Miscellaneous expense	247	. . .	247
	$15,160	$ 585	$15,745
Feed purchases	9,371	. . .	9,371
Livestock purchases	12,880	. . .	12,880
Fixed:			
Taxes	1,584	$ 297	1,881
Insurance	334	53	387
Interest (3.4)	3,703	. . .	3,703
Rent	2,000	. . .	. . .
	$ 7,621	$ 350	$ 5,971
	$45,032	$ 935	$43,967
Noncash Expenses			
Depreciation:			
Machinery	$ 3,734	. . .	$ 3,734
Building	877	$ 245	1,122
Inventory decrease:			
Crops, feed, supplies	0	. . .	0
Livestock	0	. . .	0
	$ 4,611	$ 245	$ 4,856
Total Business Debits			
Cash expenses	$45,032	$ 935	$43,967
Plus noncash expenses	4,611	245	4,856
	$49,643	$1,180	$48,823

TABLE 5.5. (continued)

Item	Operator	Landlord	Total farm
GROSS RETURN MEASURES			
Gross Profit			
Total business credits	$64,346	$4,763	$67,109
Minus:			
Inventory decrease:			
Crops, feed, and supplies	0	. . .	0
Livestock	0	. . .	0
Livestock purchases	12,880	. . .	12,880
Feed purchases (5.6)	9,371	. . .	9,371
	$42,095	$4,763	$44,858
NET RETURN MEASURES			
Net Farm Income			
Total business credits	$64,346	$4,763	$67,109
Minus total business debits	49,643	1,180	48,823
	$14,703	$3,583	$18,286
Net Farm Cash-flow Income			
Total cash income	$59,757		
Minus total cash expenses	45,032		
	$14,725		
Plus:			
New loans	13,000		
Unpaid accounts adjustment			
(ending-beginning) (3.3)	264		
Interest adjustment:			
(accrual-cash) (3.4)	70		
	$12,806		
Minus:			
Purchases (cash paid down):			
machinery and equipment	4400		
buildings and improvements	0		
Principal payments on loans	10,357		
	$14,757		
	$12,774		

*Numbers in parentheses indicate form numbers where figures are developed.

efficiency of the business. This general type of record may be useful for farmers who produce and market only a limited number and amount of products. Few U.S. farmers fall into this category.

Multicategory accounts are similar in format to the journal account except that receipts and expenses are recorded in separate accounts or parts of the book and columns are provided for dividing receipts and expenses into categories for tabulation identification. An example of a multicategory expense account is shown in Form 5.4. The receipt account is similarly constructed.*

The principal merit of the multicategory account is that it provides a simple form for recording business transactions while separating major categories. Its principal drawback is that there are limited provisions for recording physical quantities. If quantities are recorded, they are usually shown in one column. Also, the width of the account usually is so great that following a row across two pages is difficult. To add quantity columns would make the form unmanageable.

The multicategory account format is more useful for recording expenses than

*An account book of this type with wide acceptance in the Midwest is Herb Howell, Better Farm Accounting, 3rd ed.. Iowa State Univ. Press, Ames, 1972.

TABLE 5.6. Income statement tabulations illustrating differences between net cash-flow income, cash net income, accrual net income, and gross profit.

Item	Amount	Net cash-flow income	Net farm income Cash method	Net farm income Accrual method	Gross profit
Receipts					
Butcher hog sales	$18,000	$18,000	$18,000	$18,000	$18,000
Breeding swine sales	3,000	3,000	3,000	3,000	3,000
Soybean sales (total crop)	5,000	5,000	5,000	5,000	5,000
Miscellaneous farm income	1,000	1,000	1,000	1,000	1,000
Value of farm products used in home	200		200	200	200
Capital asset sales, gain	100	100	100	100	100
Undepreciated balance	200	200			
Bank loan for farm operation	5,000	5,000			
Expenditures					
Cash operating expenses	6,000	−6,000	−6,000	−6,000	
Feed purchases, all on account	3,500			−3,500	−3,500
Feed account payments	3,000	−3,000	−3,000		
Swine purchases (breeding)	600	−600	−600	−600	−600
Taxes and insurance	2,500	−2,500	−2,500	−2,500	
Interest on loans:					
Accrued amount	1,200			−1,200	
Cash payments	1,000	−1,000	−1,000		
Loan principal payments (all)	9,000	−9,000			
Depreciation:					
Machinery	2,500		−2,500	−2,500	
Improvements	500		−500	−500	
New tractor purchase					
($2,000 down, $4,000 credit)	6,000	−2,000			

Inventory	Beginning	Ending	Change			
Feeder pigs	$ 7,000	$ 7,500	$ +500		500	500
Breeding swine	2,500	2.700	+200		200	200
Feed	1,200	900	−300		−300	−300
Seed and supplies	300	400	100		100	100
Machinery	12,000	15,300	3,300			
Improvements	4,000	3,500	−500			
Land	80,000	80,000	0			

Value of Grain Production			$10,000			
Balance				$ 8,200	$11,200 / $11,000	$23,700

receipts. Quantity and price information is of greater value in the business analysis for receipts than for expenses. Also, quantity information for critical expense categories can be sorted from the description detail or the amount column.

Single-category account books use a separate page for each major category of receipts and expenses. This is much the same as having a separate journal account for each category and allows keeping much more detail surrounding each transaction, e.g., price, quantity, and enterprise or activity associations. Thus its major merit is in furnishing production and marketing detail for farm analysis purposes. Examples of single-category accounts for receipt and expense transactions are shown in Forms 5.5 through 5.8.*

*A good example of a single-category account book is the Midwest Farm Account Book by James and Trede, Iowa State University Bookstore.

Form 5.3. Receipt and Expense Journal Account

Date	Description of transaction	Receipts		Expenditures		Bank balance	
		$	xxx	$	xxx	$	869.46
	Balance forward						
Jan. 2	Sold 10 hogs to Swift's		385.31				1254.77
Jan. 2	Paid feed bill at Coop				117.98		1136.89
Jan. 2	Cashed check for cash				35.00		1101.89
Jan. 7	Purchased nails on account for farm repair				(1.50)		1101.89
Jan. 7	Purchased hammer and small tools on account				(6.35)		1101.89
Jan. 10	Paid bank loan (interest $20.00, principal $100.00)				120.00		1181.89
Jan. 15	Paid John Taylor - wages				150.00		931.89
Jan. 20	Received from Betts Miller for shelling corn		23.00				954.89
Jan. 20	Purchased feed on account				(86.72)		1035.09
January	Cash Totals	$	862.46	$	696.83	$	1035.09

Form 5.4. Cash Farm Expenses, Multicategory Account

Side A

Date	Details of transaction (kind, to whom paid, etc.)	Quantity	Total cash amount 1	Livestock expense		Auto expense 4	Truck and machine hire 5	Fuel and oil 6
				Feed 2	Other expense 3			
1-1	Coop Elevator Co.	4000	142 34	142 34				
1-3	Coop Supply Co.		1 53					
1-3	Gate Taylor	10 hrs	25 00					
1-7	General Hardware		2 25					
1-5	Petro Fuel Co.	500 gal	155 00					155 00
1-15	Harmons—shelling	2000 bu	60 00				60 00	
1-16	United Oil Co.	bottles	21 00			21 00		
	etc.							
	All other		696 69	151 96	8 15	43 90	0 00	30 00
	Totals (forward to next page)		1105 81	294 30	8 15	64 90	60 00	185 00

Form 5.4. (continued)

Side B

Machinery expense 7	Hired labor 8	Crop expense — Fertilizer and lime 9	Crop expense — Other 10	Building and fence repair 11	Electricity and telephone 12	Taxes, rent, insurance, interest, etc. 13	Misc. 14	Purchases of — Livestock 15	Purchases of — Machinery and improvements 16
2 25	25 00			1 53					
19 05	0 00	0 00	0 00	7 42	23 48	0 00	12 73	0 00	400 00
21 30	25 00	0 00	0 00	8 95	23 48	0 00	12 73	0 00	400 00

83

Form 5.5. Swine Sales, Single-Category Account, Dunn Farm

Separate sales under the categories shown to facilitate the calculation of taxable income.

		Sale of swine purchases for resale				Sale of raised stock			Sale of breeding stock		
Date	Description	No.	Total weight	Operator amount	Purchase cost	No.	Total weight	Operator amount	No.	Total weight	Operator amount
1	1-20 Iowa Pork	21	4415	$926 10	$315 00			$	2	642	$ 89 88
2	3-2 Best Pork					43	9466	2076 90			
3	3-17 Iowa Best					21	4725	907 75			
4	5-6 Best Pork	15	9465	623 70	240 00				5	1815	235 95
5	5-10 Iowa Pork					17	3848	692 64			
6	etc.										
7											
8											
9											
10											
11											
12											
13											
31											
32											
33	All other	49	11,500	2026 41	725 00	585	136,969	20899 14	46	16,517	2330 53
34	Column Totals	85	19,380	$3576 21	$1280 00	666	155,001	$24576 43	53	18,974	$2656 36
									804	193,356	30,809 00

84

Form 5.6. Feed Purchases, Single-Category Account, Dunn Farm

Entries may be made in the total columns only or by enterprise or both. "Unallocated feed" is that which cannot be listed by enterprise at the time of purchase. Feedstuffs of the same type may be grouped in the columns or notated for grouping at the end of the year.

	Date	Description	Beef feed purchased Type	Qty.	Amount	Swine feed purchased Type	Qty.	Amount	Unallocated feed Type	Qty.	Amount
1	1-10	Coop Mills			$	Brower	1500	$ 75 00			$
2	1-15	Coop Mills				Finisher	4000	194 21			
3	1-21	Herb Jenkins							Shelled Corn	784	902 00
4	1-22	Brown Fast Feeds	32%	6000	288 53	Starter	400	28 40			
5	1-28	Coop Mills				Finisher	3000	144 61			
6	2-1	Coop Mills				Brower	2000	105 17			
7	2-6	Brown Fast Feeds	32%	5000	247 33						
8	2-8	Oscar Mann							Oats	729	474 00
9											
10											
11											
12											
13											
14											
31											
32											
33		all other		19,300	912 14		94,700	5999 61			1376 00
34		Column Totals	xxx	#29,300	1448 00	xxx	#105,600	6547 00	xxx	#67,232	$ 9371 00

Form 5.7. Crop Expenses, Single-Category Account, Dunn Farm

Includes seeds, plants, fertilizers, lime, insecticide, herbicides, soil testing, crop insurance, and supplies such as baler twine and sacks. Soil and water conservation expenses may be recorded here or under capital improvements, depending on whether the improvements are capitalized.

| | Date | Description | Qty. | Operator amount | | | | | Landlord amount |
				Hay	Corn	Soybeans	Oats	Other	
1	Feb 15	Dist Seed testing		$	$	$	$	$ 35.00	$ 35.00
2	Mar 5	Ace Seed Co.	35 bu		525.00				
3	Mar 15	Grow-Glue Fert. Co.			1406.00	233.00	212.00		400.00
4	Mar 20	ASC premiums						44.00	
5	Mar 31	Chemical Products Co.				148.00			
6	Apr 10	Seiharvest supply twine		170.00					
7		.Tc'.							

| 33 | All other | | | | | | | | |
| 34 | Column Totals | xxx | $ 430.00 | $ 3128.00 | $ 514.00 | $ 374.00 | $ 102.00 | $ 525.00 |

Fertilizer 2567.00   Other 1976.00

86

Form 5.8. Monthly Labor Hired, Single-Category Account, Dunn Farm

Includes labor that requires the payment of social security (FICA) taxes. Space is provided for the employer to record his share of FICA tax and the date paid. Social security payments must be made for all employees working twenty (20) or more days or earning $150.00 or more over the taxable year.

Name of worker *Mark A. Little* Social security no. *123-45-6789*

Address *Overthehill, Iowa* Nature of work: *General farm work*

	Date	Hours, (days), months	Wage rate	Gross pay	FICA	Deductions *Insurance*		Net pay	Operator share Amount	Date
1	*Jan 31*	25	$400	$ 400.00	$ 20.80	$ 10.00	$	$ 369.20	$	
2	*Mar 1*	23	400	400.00	20.80	10.00		369.20		
3	*Mar 31*	26	400	400.00	20.80	10.00		369.20		
4		*etc.*								
5										
6										

13										
14	Column Totals			$4000.00	$ 208.00	$ 100.00	$	$ 3692.00	$ 208.00	xxx

The major disadvantage of the single-category account is in the recording of transactions and cash control. It is necessary to turn to a different page to record each separate category of receipt or expense. This is time consuming even if the account book is well organized. However, the added information may be worth the extra time required. Keeping track of cash balances is a major effort and would require a summary form to bring the monthly transactions for the separate categories together.

Many account books combine the features of the multicategory account with the single-category account. This combination allows keeping detailed information on certain items while reducing the amount of page turning required for the majority of business transaction recordings.

In summary, the following recommendations are made concerning the selection of a method for recording data in the single-entry account:

1. The journal account is most useful for the farmer who markets only a small volume and has few expenses.
2. The multicategory account is most useful for the farmer who wants financial accounts by category, such as required for tax reporting, but is not going to submit his records to detailed farm analyses.
3. The single-category account is most useful for the farmer who wants to identify receipts and expenses by enterprise or activity and is going to make detailed farm analyses.

DIFFICULT BOOKKEEPING TRANSACTIONS

Credit and debit transactions that may be problems to the bookkeeper are discussed in this section. Regular sales of crops, livestock, and livestock products; standard purchases of operating supplies, feed, and livestock; fixed expenses for taxes and interest, and similar transactions will not be given special attention. However, these should be recorded in enough detail to allow for business analyses. Quantities and weights are important for tabulating many efficiency measures and should be recorded along with the amounts paid or received.

Changes in inventory and purchases and sales of capital assets are not always handled the same in different account books or even within the same account book for different items. Sometimes total inventories are shown instead of changes in inventory. The beginning inventories are shown on the debits side and closing inventories are shown on the credits side. This does not change the net income but does give different subtotals, e.g., total business credits and total business debits.

Perhaps more perplexing is the handling of depreciation. Some systems show beginning and closing inventories and purchases and sales for depreciable properties instead of depreciation and profits or losses from sales as illustrated in Tables 5.4 and 5.5. Either method is correct and gives the same net income. Some of the subtotals will be different, however. Table 5.7 illustrates the similarities and differences in the two methods. Either the method on the left of the line or the one on the right can be used. Note that each gives the same result. The method used in Tables 5.4 and 5.5 for machinery and improvements is illustrated on the right where only depreciation and gains or losses from sales are shown. Nondepreciable assets such as crops and feeder livestock are illustrated on the left side. Table 5.8 illustrates these two methods with an actual example. Notice the handling of depreciable assets in the receipts and expense

TABLE 5.7. Net income aspects of the inventory account with respect to depreciation, purchases, and sales

Transaction or item	Beginning inventory (debits)	Purchases (debits)	Sales (credits)	Closing inventory (credits)	Depreciation (debits)	Loss from sale (debits)	Gain from sale (credits)
Depreciable asset carried over	$1,000	...	...	$ 900	$100	...	...
Depreciable asset purchased at end of year	...	$ 500	...	500	0	...	...
Depreciable asset purchased during year	...	2,200	...	2,000	200	...	...
Depreciable asset sold at a loss	300	...	$250	...	...	$50	...
Depreciable asset sold at a gain	100	...	200	...	...	...	$100
Depreciable asset purchased with a trade involved*	500	(2,300) 1,800	(700)	1,900	400	...	(200)
	$1,900	$4,500	$450	$5,300	$700	$50	$100
		$6,400		$5,750		$750	$100
			$650 Net Debit			$650 Net Debit	

*The item purchased had a $2,500 retail value. The farmer paid $1,800 cash plus the trade-in valued at $700. Since the trade-in value was $200 more than the beginning inventory value, the purchase price (tax basis) is $200 less than the retail value ($2,300). There is no sales or gain shown separately.

sections. The right-hand version where depreciation is shown as a direct expense gives less distorted subtotals, lends itself more completely to the analysis process, and reflects IRS specifications.

Livestock that are on depreciation should be treated with other livestock not on depreciation rather than being split-listed. This is done by using the depreciation schedule as a source for the beginning and ending inventory values. These values are then added to those for livestock not on depreciation.

The handling of inventory changes, including depreciation, can be understood more completely by studying how these transactions affect the net worth statement of the business in the double-entry accounts. These are illustrated in the last section of this chapter.

Capital gains or losses create another problem that is directly related to the valuation methods used in the inventory. Capital gains or losses arise from changes in value of capital assets carried over in the inventory from one year to the next. Generally speaking this refers to working and fixed assets. Depreciation has already been discussed as a method of adjusting for changes in value of depreciable working and fixed assets. Also, as old items wear out and are replaced, the general level of prices is reflected in the replacement costs. There is a lag since all items are not replaced each year, but over a period of 5-10 years a gradual adjustment takes place. The most critical items are land and breeding livestock, thus the balance of the discussion of capital gains will emphasize these, and any relevance to other items can be extrapolated.

Land values change more as a result of the general price level of the economy, and agriculture in particular, than from its productivity. Certain aspects of land cannot be destroyed and may even become more valuable over time due to increased productivity or outside economic changes. The question here is not whether land values change but whether change, if it occurs, should be reflected in the income statement.

Since the farmer is not a real estate agent, he generally holds land for the principal

TABLE 5.8. Net farm income calculated under two different methods of handling depreciable property expenses

Inventory method		Depreciation method	
Receipts		*Receipts*	
Livestock sales	$15,400	Livestock sales	$15,400
Livestock product sales	1,000	Livestock product sales	1,000
Crop sales	12,800	Crop sales	12,800
Miscellaneous receipts	400	Miscellaneous receipts	400
		Profit from depreciable asset sales	100
Sales of depreciable assets	600	Inventory increase:	
		Livestock	3,000
Inventory increase:		Feed, seed, and supplies	. . .
Depreciable assets	1,400	Value of farm products used in home	300
Livestock	3,000		$33,000
Feed, seed, and supplies	. . .		
Value of farm products used in home	300		
	$34,900		
Expenses		*Expenses*	
Operating:		Operating:	
Livestock excluding feed	$ 300	Livestock excluding feed	$ 300
Crop	800	Crop	800
Machinery and equipment repair	2,600	Machinery and equipment repair	2,600
Building repair	200	Building repair	200
Hired labor	400	Hired labor	400
Utilities	600	Utilities	600
Miscellaneous	100	Miscellaneous	100
	$ 5,000		$ 5,000
Feed purchases	$ 2,400	Feed purchases	$ 2,400
Livestock purchases	$ 8,500	Livestock purchases	$ 8,500
Fixed:		Fixed:	
Taxes	$ 1,200	Taxes	$ 1,200
Insurance	300	Insurance	300
Interest	( )	Interest	( )
	$ 1,500		$ 1,500
Purchases of depreciable assets	$ 3,000	Losses from depreciable asset sales	. . .
		Inventory decrease:	
Inventory decrease:		Livestock	. . .
Livestock	. . .	Feed, seed, and supplies	$ 2,000
Feed, seed, and supplies	2,000	Depreciation:	
Buildings	1,000	Buildings	1,000
Machinery	. . .	Machinery	1,100
	$ 6,000		$ 4,100
	$23,400		$21,500
Net Farm Income	$11,500	Net Farm Income	$11,500

purpose of using it as a factor of production. He may exchange land (buy or sell) only once or twice in a lifetime. Thus any gains or losses from holding land, aside from its productive services reflected in sale of products, cannot be realized until the land is sold. They are only on paper until this time. These value changes, however, may be useful in assessing the credit base of the farmer. Land value changes should be recognized, but probably not in the income statement. It is better accounting procedure to reflect these capital changes between the end of one accounting period and the beginning of the next. Even here it would not seem desirable or necessary to make this change every year. Perhaps an adjustment every 3-5 years would be sufficient.

The case of livestock is not so easily dealt with, although some of the same reasoning applies. For the farmer or rancher who holds breeding stock as a major enterprise, any change in value due to price level changes are paper gains or losses. For example, the farmer with breeding cows who values them in his inventory by the market price method would have inventory gains while prices were increasing and losses when they were decreasing, *but* his bank account would not reflect any change as a result of the values placed on the breeding herd. Thus there are good reasons for holding breeding livestock prices at a near-constant level over time and adjusting them only when the cost of production changes considerably. This is truer of raised breeding stock than purchased ones and truer for cattle than for sheep or swine. Purchased breeding livestock can be placed on depreciation and their value adjusted according to the method selected. Cattle are longer lived than sheep or swine, thus the length of their price swings is also greater. A sow whose price has increased may be marketed before the price falls whereas this is generally not true of a cow.

The IRS allowance for capital gains complicates the process of valuing breeding livestock as related to the income aspects of changing inventory values. Qualified capital gains are not subject to the same tax treatment as ordinary gains. Only *one-half* of capital gains are counted as income for tax purposes. Losses receive no such special treatment. Livestock (cattle, swine, horses, and sheep, but not poultry) if held for a specified number of months for draft, dairy, or breeding purposes are subject to capital gains provisions.* The total sale price of qualified raised breeding livestock would be subject to capital gains provisions for the farmer reporting income by the cash method. For the farmer reporting income by the accrual method the capital gain would be only the difference between the sale price and the inventory value at the beginning of the year. For both farmers (cash or accrual) capital gains for livestock on depreciation would be the difference between the sale prices and undepreciated values at the time of sale. Thus it is easy to see that the larger the difference between sale and inventory values the less tax the farmer will need to pay. The farmer paying taxes on the cash method has a considerable advantage where the sale of large numbers of breeding livestock is involved. The farmer reporting on the accrual basis may have a tendency to place a low value on his animals so that the size of the gain may be as large as possible.

The "cost or market, whichever is lower" and the unit-livestock-price method have much to recommend them for livestock valuation for the farmer using the accrual method of calculating his income tax. The use of these methods would make it unnecessary to keep two separate accounts—one for tax purposes and one for farm analysis. The breeding stock inventory would not be subject to the large changes due to fluctuating prices. The farmer who reports his income on the cash basis but keeps a separate record for business analyses can be more selective in choosing a livestock valuation method.

Repair versus investment is another item that may be difficult to determine. Examples would be whether to include the total cost of overhauling a tractor or reroofing

---

*For cattle and horses acquired after 1969 the holding period is 24 months, for other livestock it is 12 months. Internal Revenue Service, Farmer's Tax Guide, 1974.

the barn as expense for the year in which the expenditure occurred. Two tests may be helpful here:

1. If the expenditure increases the life expectancy of the item being repaired, it generally should be considered an investment.
2. If the expenditure has residual value that can be recovered in sale at the end of the year, it generally should be considered an investment and added to the depreciation schedule.

One should be cautious about including a cost for the operator's labor when making repairs or adding new facilities that will later be added to the depreciation schedule. If the operator's labor is included, this value would be reflected later in the depreciation expense. Thus if the operator adds his labor, he also must show a labor income (but this is circular accounting). The same is true for hired labor, particularly monthly labor used for other farm tasks. If hired labor is included in the cost of a capital improvement and also charged as a general farm expense, this is double-counting. Where farm labor is used for repair or construction of items to be added to the depreciation schedule, the operator's labor should be excluded; if the regular hired man contributes significantly, his regular wage entry should be reduced accordingly and that amount added to the repair cost or value of the new capital item to be later counted as an expense through depreciation.

Perquisites furnished hired labor should not be counted at their market value unless this reflects their cost of production. Perquisites include living quarters, food, clothing, insurance, etc. These are valid expenses and are as real as the direct salary paid. However, if they are valued above their actual cost, there is a margin of profit and this also should be reflected. Suppose the farmer's wife furnishes board, room, and laundry to the hired man. If she includes a wage for her labor when calculating the salary adjustment or charge, it should be recorded as income. This assumes the farmer and his wife jointly operate the farm. When calculating cost of living expenses, it is appropriate to include an allowance for the wear and tear on the home, cooking utensils, bedding, etc., if these are not accounted for elsewhere.

Farm-family shared expenses should be divided according to the proportion used by each, the home and the farm. This may include the telephone, electricity, gas, family auto, and newspaper. There is nothing magic about 50 or 25% except that they are easy to apply. The point is that some cost items are jointly used and a division needs to be made. Be as realistic and objective as possible in making this division. The farmer should not overlook the fact that his home often furnishes a farm business office, and the business should be so charged. The same is true of the garage that shelters the farm truck, farm-shared auto, or farm supplies from time to time.

Farm products used in the home should be included in income where a business expense has been charged for their production. However, they should be valued at no more than is reflected in the expense account for their production. Otherwise a profit is allowed. An alternative is to not include their cost as a business expense. The IRS prefers the latter.

Work performed off the farm for hire should be counted with farm income until it is considered a major separate business. This includes custom machinery work as well as labor. Separate accounts are not justified until it is desirable for efficiencies of production for the off-farm work to be calculated. Neighborly exchanges do not usually justify separate accounts.

Crop commodity loans can be handled in two ways. (1) The loan can be counted as a sale. When the crop is later sold, any differences between the loan value and sale amount can be recorded. (2) The loan can be counted as a money loan until the crop is actually sold. In the latter case, the amount of crop involved would be included in the closing inventory if it ended between the time of the loan and sale of the crop. Either method is allowed by the IRS, providing the method selected is used in all future accounting periods. Treating the amount received as a money loan would seem to have merit for proper accounting procedures.

Accrued expenses which have not been paid should be recorded in the year incurred rather than the year paid. Examples might be accrued interest on a loan whose payment does not fall due at the end of the accounting year, taxes which are paid the following year, and wages earned by an employee but not paid. These take the form of an increased liability (credit) and a decreased owner's equity.

Prepaid expenses are similar but opposite to accrued expenses. In this case a payment has been made covering an expense that has not yet occurred. Examples are interest paid at the beginning of a loan, insurance premiums paid in advance, wages paid in advance to employees, and fertilizer purchased in advance of application. These expenses should be prorated on the basis of the accounting year when the service is rendered.

Principal payments on liabilities are *not* business expenses. An expense was incurred when the borrowed money was used for business purposes. Principal payments are in a sense an investment. When a principal payment is made, only the net worth statement is involved, not the income statement. Either the cash account is reduced or another liability is incurred. However, there is some justification for including principal payments as cash expenditures when calculating net cash-flow income. Only the cash paid down on new purchases would be recorded as cash flow. The new cash-flow income then would give the net cash available for family living and outside investments. This is a useful figure.

Interest payments are considered as business expenses in most accounting systems. Certainly they are an expense to the farm operator or owner. However, in some accounting systems interest is considered a return on the capital invested and is not treated as an expense. This has some logic if the farm is looked at in the corporate sense as an individual. The farm is comprised of its resources, which are listed as assets in the net worth statement. The liabilities, including the operator's net worth, are claims upon these resources. Interest is a payment for the use of capital of another, in this case the liability claims upon the assets. Thus interest is a business expense of the operator. On the other hand, interest can be looked upon as a return to the business assets. A rental payment is also of this same nature; hence it is logical to consider interest as a return to the claimants of these resources. Most outside liabilities are covered by security interest agreements and real property mortgages. This book considers interest as a business expense. This is why net farm income was defined to include ''a return to unpaid capital.'' When analyzing the farm business records and accounts, interest will be subtracted from net farm income and treated as a return.

There are probably some other areas not covered by this discussion that will be troublesome to some farm accountants or readers. It is hoped that the reasoning used in the various items treated will be a useful carry-over. A study of the double-entry accounting procedures in the last section of this chapter will help show how a business transaction affects the various accounts.

PROCEDURES IN ADDING A CASH JOURNAL

Under normal single-entry accounting systems a certain amount of control over receipts and expenditures is lost. Also, it is difficult to reconcile the cash balance with the receipt and expenditure accounts. It can only be assumed that all transactions are recorded somewhere in the accounts and all will balance at the end of the accounting period. Complicating this more is the fact that most farmers do not maintain a separate cash or checking account for the farm, aside from the family or outside investments. Still another important factor is the deferred payment. The problem is ever present of how to handle purchases or receipts where the item is purchased or sold but the payment is still pending. These omissions and confusions may have serious consequences resulting in increased tax liability and inaccurate farm analyses.

If the above poses a serious problem in your situation, it may be possible to remedy most of these difficulties by adding a cash account or journal to the present single-entry system. This addition carries with it some of the added work of the double-entry system, but the control may be worth the time and effort involved.

The cash account can be added separately or as a part of the present single-entry system where a multicolumn expense record is used. In order to better present the principles involved, the separate cash journal will be treated first. It will be assumed that a joint farm and personal account is maintained. The principles and procedures would be similar if separate accounts were kept.

Every receipt or expenditure transaction involves the cash account, which consists of the bank balance and perhaps the in-pocket money. The latter need not be included if most farm expenses are paid by check. Thus if a running balance between receipts, expenditures, and the cash account is maintained, greater control can be exercised. The cash account is increased with each receipt and decreased with each expenditure. Receipts may be from the sale of farm goods, outside investments, nonfarm labor, gifts, etc.,as well as incurred liabilities in the form of floating bills, notes, and mortgages. Expenditures may be for farm expenses, family purchases, and outside investments. Since outside investments are of a minor nature on most farms, they will be considered here with the personal receipts and expenses. It is convenient to check the balance in the cash account at the time the monthly bank statement is received. Many individuals make most of their payments and recordings on a monthly basis, so this would fit conveniently into their present system.

Form 5.9 illustrates how to develop a cash journal and how transactions are to be recorded. For every transaction listed in the farm expenditure column, a duplicate listing is entered in the regular farm account book. A posting reference column may be used to record the section to which the entry is transferred. If a personal account is maintained, a similar reference can be made. It will be noted that farm expenditures, whether for cash or credit, are listed only once. Any principal payments on farm credit accounts are then entered in the "nonfarm cash" column since liabilities are of a personal nature even if incurred against the farm business. At the end of each month the

Form 5.9. Example Entries for a Farm and Nonfarm Cash Journal

| Date (1974) | Description | Receipts | | | | Expenditures | | | | Cash balance | Posting reference |
		Nonfarm Cash	Nonfarm Deferred	Farm Cash	Farm Deferred	Nonfarm Cash	Nonfarm Deferred	Farm Cash	Farm Deferred		
Mar. 3	Purchased TV set for $275 with a $25 down payment	$	$	$	$	$ 25	$ 250	$	$	$1,025 1,000	Family and credit
Mar. 13	Purchased fertilizer for farm							150		850	Farm
Mar. 20	Purchased livestock feed on open account from Valley Coop								100[pd]	850	Farm and credit
Mar. 20	Purchased family groceries					20				830	Family
Mar. 30	Made payment on TV					25				805	Credit
Apr. 19	Sold cattle at auction			1,500						2,305	Farm
Apr. 19	Made feed payment to Coop					100				2,205	Credit
May 5	Obtained farm bank loan	1,000								3,205	Credit
May 5	Purchased tractor for $3,000 with $1,000 down							1,000	2,000	2,205	Capital and credit
May 10	Paid farm share electric bill					15		15		2,175	Farm and family
June 3	Made payment on bank loan--$500 principal, $20 interest					500		20		1,655	Credit and farm
June 4	Received notice of interest credited to nonfarm investment		25							1,655	Deferred
June 5	Made tractor payment					100				1,555	Credit
June 11	Sold sow, to pay next month				60[pd]					1,555	Farm and deferred
June 13	Coop patronage certificate received								50	1,555	Deferred
June 15	Cashed check for pocket money					25				1,530	None
June 15	Purchased gate hardware with cash							10		1,520	Farm
June 17	Received payment for sow	60								1,580	Deferred

bank balance can be compared with the cash balance. Any errors can then be discovered and corrected.

It is possible to hold posting of transactions to the farm account or personal account until the end of the month or some other convenient period. Reference columns can be added to show when and where these postings are made. A hold file could be maintained for the interim.

Much flexibility exists in the use of a cash account. Where separate family and farm bank accounts are maintained, the nonfarm expenditure columns are no longer necessary. However, it would be necessary to show transfers from the farm account to the family account and vice versa. Also, a slight change would be required on the farm cash account. In the method illustrated, principal payments on previously incurred liabilities were entered in the nonfarm cash column. Where the accounts are separated, it would be normal practice to make all payments on farm liabilities from the farm cash account. A possible method for differentiating between principal loan payments and other expenses is to place brackets around the loan payment entries as in Form 5.10. The total of the "Cash" and "Deferred" columns, excluding the bracketed entries, is now equal to the total farm expenditures. An alternative to the above procedure of bracketing principal loan payments would be to add another column designated "liability payments," or some similar title.

Where it does not seem useful to maintain a "cash balance" column this could be dropped from the cash account also. In this case, the cash expenditures are totaled at the end of each month or balancing period and deducted from the total monthly receipts (bank deposits) to determine changes that have occurred. The net change can be checked against the bank statement for possible errors.

If a separate cash account cannot be justified, it may still be desirable to add "cash payment" and "deferred payment" columns to the regular multicolumn expense account. In this case, every expense transaction first would be entered in the cash or deferred column prior to its entry under a specific category of expense. Where all expense categories are not found on the multicolumn listing, a reference column could be maintained. Form 5.10 illustrates how this account could be set up.

The cash journal just discussed somewhat resembles the cash-flow statement. This is used by some accountants (particularly those using the checkbook method) to show the flow of cash out of the bank account. Another form is found in the cash-flow plan developed by credit institutions when planning farm loan needs.

DOUBLE-ENTRY ACCOUNTING PROCEDURES

The double-entry system grew out of the fundamental balance of assets and the claims upon them discussed in connection with the net worth statement. Double-entry accounting simply involves keeping track of all changes in assets and the claims upon those assets each time a transaction is made. This is accomplished through a system of credits and debits. Each credit transaction must be balanced by a debit transaction and vice versa. Credits are defined as:

1. Increase in owner's equity (net worth)
2. Increase in a liability
3. Decrease in an asset

Form 5.10. Example Entries Where Cash and Deferred Payment Columns Have Been Added to a Multi-column Expense Account

Date	Description	Cash payment	Deferred payment	Farm account	Truck and machine hire	Telephone and electricity	Insurance and taxes	Misc. supplies	Dues, magazines, misc.	Etc.
3/13	Fertilizer	$ 150	$	Crops	$	$	$	$	$	$
3/20	Feed		100	Loan Feed						
4/19	Feed payment	[100]		Loan						
5/5	Tractor	1,000	2,000	Loan Capital						
5/10	Electric	15				15				
6/3	Bank payment	[520]		Loan						
6/5	Tractor payment	[100]		Loan						
6/15	Hardware for gate	10						10		

Note: An alternative to this method where liability payments are bracketed is illustrated in Emery N. Castle, _Farm Business Management_, p. 53, Macmillan, 1972.

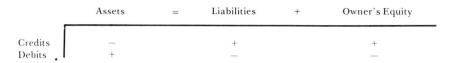

Fig. 5.1. Definition of debits and credits.

Debits are defined as:

1. Decrease in owner's equity (net worth)
2. Decrease in a liability
3. Increase in an asset

These definitions of credits and debits are illustrated in Figure 5.1. This system as defined may seem unreasonable if the objectives of farm ownership are not kept clearly in mind. One of the major objectives is to increase the net worth (equity) of the operator. Thus any increase in owner's equity is a credit and every decrease is a debit. If we begin here with our logic, the pattern fits reasonably well together as illustrated in Figure 5.2. Keep in mind that the value of assets must be equal to the claims upon them at all times, and the sum of credit entries must be equal to the sum of debit entries for every transaction. Suppose an asset is added:

Fig. 5.2. Relationship between debits and credits and the net worth statement.

1. If it comes from production on the farm:
   a. The asset account receiving the new product would be increased (debited).
   b. The owner has a first claim to it, and it would be credited appropriately in the owner's equity account.
2. If it comes from outside the business:
   a. If it is a cash purchase, (1) the capital asset account receiving the item would be increased (debited) and (2) the cash asset account would be decreased (credited).
   b. If it is a deferred payment purchase, (1) the capital asset account receiving the item would be increased (debited) and (2) the liability account would be increased (credited).

Double-entry ledgers are organized as in Figure 5.3. The debits are always recorded in the left column and the credits on the right.

Single-entry accounting is most concerned with the owner's equity accounts in the context of the general formula for double-entry accounting. Referring again to debits and credits it can be seen that a credit to owner's equity is caused by a receipt (increase) and a debit is caused by an expense (decrease). Single-entry accountants concern

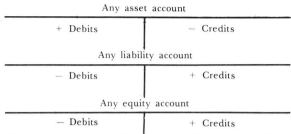

Fig. 5.3. Organization of a double-entry ledger.

themselves primarily with the receipts and expenses of the business without any real effort to maintain a balance in the basic accounting formula. This does not mean that asset accounts (capital purchases and sales, depreciation schedules, etc.) and liability accounts (credit) are not maintained but that the balance of credit and debit entries are not.

Consider the effect of the following business events upon the net worth statement. These are illustrated in Examples 5.1-5.13 and summarized in Table 5.9.

1. A new machine is purchased.
   a. If this is from accumulated savings, the cash asset account is decreased (credited) and the machinery asset account is increased (debited). Owner's equity is not affected (Example 5.1).
   b. If this is by credit, the machinery asset account is increased (debited) and the liability account is increased (credited). Again owner's equity is not affected (Example 5.2).
   c. Only when the new machine depreciates is the owner's equity account affected. The machinery asset account is decreased (credited) and the owner's equity account is decreased (debited) by showing a business expense (Example 5.3).
   d. When a credit payment is made, the principal amount does not offset the equity account; only the interest affects the equity account (Example 5.4).
2. Farm-raised grain is sold.
   a. If this grain had not been previously recorded (inventoried), the cash-asset account would be increased (debited) and the owner's equity account would be increased (credited) (Example 5.5).
   b. If this were grain that had been inventoried (last year's crop), the grain-asset account would be decreased (credited) and the cash-asset account would be increased (debited). The owner's equity account would only be affected by the difference between the inventory value and the sale value (Example 5.6).
3. Feeder livestock are sold.
   a. If these were purchased from accumulated cash, (1) the cash account would be increased (debited) by the total amount of the sale, (2) the livestock asset account would be decreased (credited) by the inventory value of the cattle if purchased in the previous period or by their purchase price if purchased in the same accounting period, and (3) the owner's equity account would be increased (credited) by the difference be-

EXAMPLE 5.1
   Operator purchases a new tractor for $8,000 and pays cash.

	Cash-asset account	
	Debit (+)	Credit (−)
Before amount	$10,000	
Tractor purchase		$8,000
After amount	$ 2,000	

	Machinery-asset account	
	Debit (+)	Credit (−)
Before amount	$12,000	
Tractor purchase	8,000	
After amount	$20,000	

EXAMPLE 5.2
   A new plow is purchased for $1,000 and financed with dealer credit.

	Machinery-asset account	
	Debit (−)	Credit (+)
Before amount	$20,000	
Plow purchase	1,000	
After amount	$21,000	

	Liability account	
	Debit (−)	Credit (+)
Before amount		$100,000
Plow purchase		1,000
After amount		$101,000

EXAMPLE 5.3
   The operator writes off the depreciation experienced on the tractor and plow purchased in Examples 1 and 2.

	Machinery-asset account	
	Debit (+)	Credit (−)
Before amount	$21,000	
Depreciation		$900
After amount	$20,100	

	Operator's equity account	
	Debit (−)	Credit (+)
Before amount		$100,000
Depreciation	$900	
After amount		$ 99,100

EXAMPLE 5.4

The operator makes a principal payment of $500 and interest payment of $35 on a farm loan (like the dealer loan to buy the plow).

	Cash-asset account	
	Debit (+)	Credit (−)
Before amount	$2,000	
Principal and interest amount		$535
After amount	$1,465	

	Liability account	
	Debit (−)	Credit (+)
Before amount		$101,000
Principal payment	$500	
After amount		$100,500

	Equity account	
	Debit (−)	Credit (+)
Before amount		$99,100
Interest payment	$35	
After amount		$99,065

EXAMPLE 5.5

The operator harvests his oats and sells the entire crop for $3,000.

	Cash-asset account	
	Debit (+)	Credit (−)
Before amount	$1,465	
Grain sales deposit	3,000	
After amount	$4,465	

	Equity account	
	Debit (+)	Credit (−)
Before amount		$ 99,065
Grain sales deposit		3,000
After amount		$102,065

EXAMPLE 5.6

The operator sells some of last year's corn crop for $2,000. It was inventoried at $1,800 in the beginning of the year inventory.

Grain-asset account

	Debit (+)	Credit (−)
Before amount	$10,000	
Sale of grain (inventory reduction)		$1,800
After amount	$ 8,200	

Cash-asset account

	Debit (+)	Credit (−)
Before amount	$4,465	
Sale of grain	2,000	
After amount	$6,465	

Equity account

	Debit (−)	Credit (+)
Before amount		$102,065
Increase from sale of grain		200
After amount		$102,265

EXAMPLE 5.7

The operator purchases feeder cattle for $5,000 and pays cash.

Cash-asset account

	Debit (+)	Credit (−)
Before amount	$6,465	
Cattle purchase		$5,000
After amount	$1,465	

Cattle-asset account

	Debit (+)	Credit (−)
Before amount	$ 6,000	
Cattle purchase	5,000	
After amount	$11,000	

EXAMPLE 5.8

The operator sells cattle that were on inventory at $6,000 for $10,000.

	Cattle-asset account	
	Debit (+)	Credit (−)
Before amount	$11,000	
Cattle sold		$6,000
After amount	$ 5,000	

	Cash-asset account	
	Debit (+)	Credit (−)
Before amount	$ 1,465	
Cattle sold	10,000	
After amount	$11,465	

	Equity account	
	Debit (−)	Credit (+)
Before amount		$102,265
Cattle sold		4,000
After amount		$106,265

EXAMPLE 5.9

The operator takes out a bank loan for $7,000 and uses the amount to purchase feeder cattle.

	Liability account	
	Debit (−)	Credit (+)
Before amount		$100,500
New cattle loan		7,000
After amount		$107,500

	Cash-asset account	
	Debit (+)	Credit (−)
Before amount	$11,465	
New cattle loan	7,000	
Purchase of cattle		$7,000
After amount	$11,465	

	Cattle-asset account	
	Debit (+)	Credit (−)
Before amount	$ 5,000	
Purchase of cattle	7,000	
After amount	$12,000	

EXAMPLE 5.10
The operator sells cattle that were purchased for $7,000 for $13,000.

	Cattle-asset account	
	Debit (+)	Credit (−)
Before amount	$12,000	
Cattle sale		$7,000
After amount	$ 5,000	

	Cash-asset account	
	Debit (+)	Credit (−)
Before amount	$11,465	
Cattle sale	13,000	
After amount	$24,465	

	Equity account	
	Debit (−)	Credit (+)
Before amount		$106,265
Cattle sale		6,000
After amount		$112,265

EXAMPLE 5.11
The operator pays off cattle loan—$7,000 principal, $400 interest.

	Cash-asset account	
	Debit (+)	Credit (−)
Before amount	$24,465	
Principal and interest payment		$7,000
After amount	$17,465	

	Liability account	
	Debit (−)	Credit (+)
Before amount		$107,500
Loan principal payment	$7,000	
After amount		$100,500

	Equity account	
	Debit (−)	Credit (+)
Before amount		$112,265
Loan interest payment	$400	
After amount		$111,865

EXAMPLE 5.12

The operator purchases fertilizer (applied) for $1,500 and pays cash.

	Cash-asset account	
	Debit (+)	Credit (−)
Before amount	$17,465	
Purchase of fertilizer		$1,500
After amount	$15,965	

	Equity account	
	Debit (−)	Credit (+)
Before amount		$111,865
Purchase of fertilizer	$1,500	
After amount		$110,365

EXAMPLE 5.13

The operator inventories his cattle and places a value of $8,000 on them.

	Cattle-asset account	
	Debit (+)	Credit (−)
Before amount	$5,000	
Inventory increase	3,000	
After amount	$8,000	

	Equity account	
	Debit (−)	Credit (+)
Before amount		$110,365
Inventory increase		3,000
After amount		$113,365

tween the purchase price and the net amount received in sale (Examples 5.7 and 5.8).

b. If these feeder livestock were purchased with a loan, (1) the cash asset account would be debited with the amount of the sale, (2) the livestock asset account would be credited with the value the livestock had when purchased if sold in the year of purchase or with the inventory value if sold in the following accounting period, and (3) the equity account is credited with the difference. It should be noted that the same net effect is obtained if the asset account is debited with the difference and the equity account shows a debit for the purchase price and a credit for the sales amount (Examples 5.9, 5.10). When the loan is retired, (1) the cash asset account is credited with the combined amount of principal and interest payment, (2) the credit account is debited with the principal payment, and (3) the equity account is debited with the interest payment (Example 5.11).

TABLE 5.9. The effect of a business event upon the net worth statement (summary of Examples 1-13)

Transaction			Asset accounts		Liability accounts		Equity accounts	
Reference number*	Example number	Description	Debit (+)	Credit (−)	Debit (−)	Credit (+)	Debit (−)	Credit (+)
		Beginning balance	$200,000	...	...	$100,000	...	$100,000
1 a	1	Tractor purchased for cash	8,000	$ 8,000	...	...	...	...
1 b	2	Plow purchased on credit	1,000	...	...	1,000	...	...
1 c	3	Tractor and plow depreciation	...	900	...	...	$ 900	...
1 d	4	Principal ($500) and interest ($35) payment	...	535	$ 500	...	35	...
2 a	5	New grain sold	3,000	...	...	...	...	3,000
2 b	6	Old grain sold for $200 over inventory value	2,000	1,800	...	...	...	200
3 a	7	Cattle purchased for cash	5,000	5,000	...	...	...	...
	8	Livestock on inventory sold	10,000	6,000	...	...	...	4,000
3 b	9	Cattle purchased on credit	7,000	...	...	7,000	...	...
	10	Cattle purchased with credit, sold this year	13,000	7,000	7,000	...	...	6,000
	11	Cattle loan retired	...	7,400	7,000	...	400	...
4 a	12	Fertilizer purchased for cash	...	1,500	...	...	1,500	...
5 a	13	Inventoried cattle	3,000	...	...	...	...	3,000
		Transaction Balance	$ 52,000	$38,135	$7,500	$ 8,000	$2,835	$ 16,200
		Difference	$ 13,865			$ 500		$ 13,365
		Ending Balance	$213,865			$100,500		$113,365

*See text discussion.

4. Fertilizer is purchased and applied.
   a. Assuming this is a cash purchase, the cash asset account is decreased (credited) by the purchase amount and the owner's equity account is decreased (debited) in like amount (Example 5.12).
5. A productive asset is inventoried.
   a. The asset account is debited by the increase and the equity account is credited by the same amount (Example 5.13).

There is no financial transaction that does not affect the balance of assets, liabilities, and owner's equity in some way. (When properly posted, a balance of assets is maintained with liabilities plus owner's equity.) In double-entry accounting this balance is maintained throughout the accounting year and is verified by the inventory at the end. However, in single-entry accounting only the owner's equity account receives attention during the accounting year. It alone gives net income. At the close of the accounting year the income statement (owner's equity account) is used to interpret and explain the changes that have occurred in the net worth statement. The net income for both double-entry and single-entry accounting for the problem illustrated in Table 5.9 is $13,365.

When money is transferred out of the business, such as for family living purposes, the cash-asset account is reduced (credited). To maintain the accounting balance, the equity account also is reduced (debited). This equity debit is different from other business expenses such as hired labor or farm fuel. In a sense it is a withdrawal by the operator as part of his total return for the use of his labor, management, and capital in the business. Thus it is a withdrawal of his net farm income from the business. A complete accounting of the family cash withdrawals plus the increase in business equity should equal net farm income. It was reported earlier that the net cash-flow income was accounted for by the cash account balance change (bank balance increase or decrease), the cash used for family living, and other cash business investments. Similarly, the cash transfers out of the farm business for family living cover many of the cash elements of net farm income. There is a need for both business and family columns in the expense accounts of the farm business to explain all cash flows.

This brief introduction to double-entry accounting is intended only to show the similarities and differences between the two systems. The equity accounts of the double-entry system give the same net income and are similar in most respects to the receipt and expense accounts of the single-entry system. They differ in how certain entries are recorded, but not in concept. Many accounting systems are hybrids in that elements of both systems are in evidence. This is particularly true with regard to single-entry accounting. Cash balance sheets, capital accounts, credit accounts, etc., are illustrations where a transaction is recorded in more than one account.

6

PRODUCTION AND STATISTICAL RECORDS

PRODUCTION RECORDS ARE FOR RECORDING the physical performance of crop and livestock enterprises. Statistical records relate to efficiency in the use of resources such as labor, machinery, feed, and fertilizer. Whether information is production or statistical in nature will not be distinguished. The discussion will center upon the body of information useful in analyzing a particular segment, enterprise, activity, or aspect of the business. Since the number and kinds of records that may be found useful by different farmers are so numerous, only those thought to be most useful to most farmers will be discussed in general. Thus this discussion will center on crop, livestock, and labor records. Under each of these major divisions, only those records (and methods for keeping them) thought to be of greatest interest and use will be discussed.

The detail in which records are kept is a function of their intended use. There is a time expense in keeping records, and as with other farming activities the law of diminishing returns holds true. Additional records should be added only as long as anticipated benefits outvalue the costs (in dollars or other measures) of adding them. Considerable information relating to production and resource efficiency can be gleaned from the inventory, receipt, and expense accounts if they are kept in sufficient detail. This requires that numbers, weights, and other quantity measures be recorded in addition to the price and value amounts. Important measurement opportunities for determining physical production efficiencies from the financial accounts will be illustrated in this section as part of the physical and statistical records discusssion.

CROP RECORDS

Farm maps are not crop records but will be discussed here because they are very useful when developing the cropping program and keeping track of productivity and important inputs. There are many different kinds of farm maps the operator may find useful to keep. For example, he may wish to have a map of the farmstead showing the location of buildings, fences, wells, water lines, etc. This map would be of great value when planning changes in the farmstead such as adding new buildings. It is much less costly to make errors on paper by trying out a location for a new building than to construct the building only to find it is in the wrong place. Also, a map showing the legal bounds of the farm is useful for many purposes. Some of these uses will be treated later in the discussion of crop production records. Air photos aid in developing maps of the farm. These often can be obtained from the Agricultural Stabilization and Conservation Service. Maps other than those suggested above may be of general interest, but the balance of the discussion will relate to those particularly useful for crop production purposes.

All farmers producing crops should be acquainted with the soils they are farming. Soil maps are a means of gaining this familiarity. Particular attention should be paid to soil productivities, conservation needs, and other special features. Field and crop boundaries, crop selection, production practices, etc., should be developed after considering the peculiarities of the soil and differences in land form.

The Soil Conservation Service (SCS) has developed soil maps and cropping systems that conserve land use for most localities. Such a map is shown in Figure 6.1 for a small area in Polk County, Iowa, The small letters represent different soils separated by the soil boundary lines. The Clarion series, which predominate in the area, consist of well-drained soils that developed from calcareous Cary glacial till of loam texture. The slopes range from 0 to 30% and are mostly short and irregular. The surface layer is generally loam. These soils are moderately fertile and the upper layers are slightly acid. The gently sloping phases are highly productive under good management and are used intensively for row crops. The steeper slopes are used mostly for pasture. Erosion control is a serious problem on many of these soils. A representative profile of Clarion loam follows:

0-10 inches, very dark brown, friable loam
10-36 inches, dark brown to yellowish brown, slightly firm to light clay loam
36-50 inches, light yellowish brown, friable, calcareous loam

Clarion loam (0-2% slopes) can be farmed intensively; a suitable rotation is corn-soybeans-corn-oats-meadow.*

If this service is available, it is an excellent source for soils and crop information. Soil maps are often developed from air photographs, and soils can be associated easily with topographic features. Where SCS mapping services are not available, farmers may find it useful to develop their own maps showing important features such as wet spots, ditches, rocks, groves, etc., and to describe distinguishing soil differences.

Field maps are perhaps the most important of all those that farmers keep. They are useful for developing crop plans and for recording acres, varieties, treatments, particular problems, and production. These can often be developed over other farm maps such as the soil map or air photo of the farm. Some farmers make a map showing permanent boundaries of fields or other features and then use overlays upon which to show each year's crops. Most farm account books provide for developing a land-use map.

*SCS, USDA, Soil Survey of Polk County, Iowa, Series 1953, No. 9, in cooperation with Iowa Agr. Exp. Sta., Ames, June 1960.

Fig. 6.1. Soil Conservation Service map showing soil types.

Crop production records are a must even for minimum farm analyses. Crop yield information is needed not only for measuring crop efficiency but also for livestock feed determinations. (Efficiency will be treated in the analysis section, and the use of records for determining choice of livestock feed will be discussed under "livestock feed records.") In themselves, yield records are not very meaningful unless data surrounding particular yield levels are recorded. Previous crops, varieties, fertilizer rates, pesticides, peculiar weather, and other meaningful variables are all aids in interpreting production. If rather permanent field boundaries are in existence, either natural or developed, it may be convenient to record this information by field identification. Such a record could be developed for each field and several year's data could be kept in the same record. Form 6.1 illustrates one method of keeping such a record. The makeup of the form will vary with the nature of the information to be recorded. These data, as indicated earlier, could be recorded on a crop map. However, where fields are small, recording space may be insufficient to show all the desired information.

Through the study of production records and soils, varieties, fertilizer, and other inputs and variables, farmers can gain insights that will give direction to future farm plans.

A *crop summary record* is needed whether or not individual field or crop records are kept. This pulls together yields for each crop and the acres involved so that total production can be calculated. Since corn on the cob differs from shelled corn, and high-

Form 6.1. Record of Crop Production

Field No. 2
Acres 54
Soil Nicollet loam; Webster clay loam
Characteristics wet spot in SE corner

Year	Crop	Yield			Variety and rate	Fertilizer applied	Chemicals applied	Comments
		Unit	Per acre	Total				
65	soybeans	bu	31	1674	Chippewa 50,000	None	None	Weather was wet at harvest time and some beans molded.
66	corn	bu	94	5074	Hybrid 35 18,000	0-40-20 p.d. 10-20-10 start. 125-0-0 s.d.	Chlordane	Excellent spring moisture. Hot and dry in July & Aug. Small amt. of rootworm.
67	corn	bu	103	5562	Hybrid 40 20,000	10-40-20 p.d. 10-20-10 sidpt. 130-0-0 s.d.	Diazinon	Heavy spring rains. Dry but cool summer. Considerable dry rot in fall.

Form 6.2. Crop Production Summary

	Crop	Unit	Price	Field	Acres	Yield	Production	Amount	%	Qty.	Amount	Qty.	Amount
							Total farm		Landlord share			Operator share	
1	Corn for grain							$			$		$
2	Picked	bu	1.00	H 1	40	110	4400	4400 00	—			4400	4400 00
3	Combined	bu	1.07	H 2	54	103	5562	5951 00	—			5562	5951 00
4	Combined	bu	1.07	R 1	25	95	2375	2541 00	50	1187	1270 00	1187	1271 00
5	Combined	bu	1.07	R 3	15	85	1275	1364 00	50	637	682 00	637	682 00
6	Combined	bu	1.05	C 2	39	90	3510	3722 00	cash rent			3510	3722 00
7													
8	Silage												
9													
10	Soybeans	bu	2.40	H 7	12	33	396	950 00				396	950 00
11		bu	2.40	C 1	25	32	800	1920 00	cash rent			800	1920 00
12													
13	Small grains Oats	bu	.65	H 4	34	51	1734	1127 00					
14		bu	.65	R 2	27	54	1458	948 00	50	729	474 00	729	474 00
15	Straw (Oat)	45# bales	.25	H 4 + R 2	(61)		450	113 00				450	113 00
16													
17	Hay--1st cutting	ton	18.00	H 5,6	23	3.5	82	1476 00				82	1476 00
18	2nd cutting				(23)								
19	3rd cutting				( )								
20													
21													
22	Diverted acres	xxx	xxx	H 3	23.4	xxx	xxx	1490 00		xxx		xxx	1490 00
23		xx	xx	R 4	10	xx	xx	638 00	50	xx	319 00	xx	319 00
24		xx	xx	C 3	10	xx	xx	627 00	cash rent	xx		xx	627 00
25	Total crop acres	xxx	xxx	xxx	337	xxx	xxx	xxx	xxx	xxx	xxx	xxx	xxx
26	Rotated pasture	acre	9.00	H 8	6	—	—	54 00				6	54 00
27													
28													
29	Total rotated acres	xxx	xxx	xxx	343	xxx	xxx	xxx	xxx	xxx	xxx	xxx	xxx
30													
31	Improved permanent pasture	acre	5.00	H 9	30			150 00				30	150 00
32	Native improved pasture												
33													
34	Woodland	xxx	xxx	xxx		xxx	xxx		xxx	xxx		xxx	
35	Waste, etc.	xxx	xxx	xxx	13.6	xxx	xxx	xxx	xxx	xxx	xxx	xxx	xxx
36	Farmstead	xxx	xxx	xxx	13	xxx	xxx	xxx	xxx	xxx	xxx	xxx	xxx
37	Total Farm	xxx	xxx	xxx	400	xxx	xxx	$27,471 00	xxx	xxx	$ 2745 00	xxx	$24,726 00

moisture corn from dried corn, this should be reflected. Rented land should be shown separately from owned land because the landlord often shares in the crop production. The separation of shares should be made after the total production for the farm, owned and rented, has been determined. Form 6.2 shows one type of crop production summary record and illustrates how the data are recorded.

Measurement is a difficult problem that must be dealt with differently on every farm. If the crop is all sold, this is no problem. If it is stored, less accurate methods of measurement are needed. Number and volume techniques have been used. For example, if the weight of one bale of hay is known, it may be possible to determine the total weight by counting all bales and multiplying by the weight per bale. Cubic measurements can be made of the contents of silos, bins, barns, etc., and these can be converted to standard units. It is important that the unit of measure be a common one and correspond to the price unit. Conversion tables, weights and measures, and other useful measuring devices are available in many farm account books. (See Table A.1 for weights and measures.)

LIVESTOCK AND POULTRY RECORDS

Livestock, including poultry, records can be divided into four major categories: mortality, breeding, performance, and feed. Performance records can be further broken down into birth records and production records.

Mortality records are useful in keeping track of livestock numbers, but their value is enhanced if the cause of death is recorded. With this information, disease prevention plans can be developed for the future. Historical mortality figures give some idea about the percentage of death loss that can be predicted for the future. The larger the number of animals involved the better the estimate will be. Form 6.3 illustrates one type of death

Form 6.3. Mortality Record

	Date	Livestock class or description	No. died	Weight or age	Amount	Cause of death
1	Feb	Sows	2	325	$ 120 00	Lepto
2	Jun	Feeder hogs	2	60	36 00	Unknown
3	Jul	Feeder hog	1	150	25 00	Unknown
4	Aug	Heifer	1	800	180 00	Bloat
5	Nov	Feeder hog	1	175	30 00	
6	Nov	Feeder pig	1	45	15 00	Pneumonia
7	Dec	Feeder hogs	2	125	35 00	Virus
8						
9						
10						

record. Mortality information may be helpful when interpreting the income statement. If the numbers are unusual, this might cause the net income to be lower or higher.

Breeding records can be developed for individual animals or for groups of animals. Purebred producers usually desire more detail than commercial producers. Dates are useful for planning farm operations. Sires can be indicated if pedigrees are desired, or if they are used for rate of gain determinations. Form 6.4 illustrates one type of breeding record.

There are a variety of *performance or production records*. The birth record is important for recording the number of births, information relating to dam and sire, and dates useful for planning. Form 6.4 gives weaning information and implies death losses between birth and weaning times. Birth percentages and numbers can be calculated from this record. It is necessary to have a record of the number of animals in the breeding herd for these determinations. This may be taken from the inventory or from the birth record.

Weaning records can be either by numbers or weights. For the operator selling feeder animals, weight is an important aspect. Rate of gain figures relate to both the dam and the sire as well as to feed supply and related inputs.

Another set of production records relate to *products* such as milk and eggs. The amount of detail recorded can vary greatly depending upon intended use. Information can be developed for each cow, pen, or cage or more generally for a whole enterprise. Production data can be recorded daily, weekly, or at some other convenient time, again depending on its intended use and on production practices. These are commonly referred to as barn records because the record books usually are kept near the place where the production is measured. Summaries and interpretations can be made as often as the manager finds use for them—monthly, quarterly, yearly, or at the end of a production period. However, if only group production information is wanted, sales and other business accounts can often be used for production determinations. For example, if the total of eggs sold is recorded and to this is added the farm-used eggs, the total of eggs produced is determined assuming the same beginning and ending inventory. Then dividing by the total number of layers gives the average number of eggs produced per hen. It may be desirable to record the number of layers (or lactating cows) on a monthly basis in order to determine the average number of producers for the year. Values produced per head can be calculated the same way.

Meat animal production records can be developed over a wide range of situations. Records of growth can be kept for individual animals, lots or pens, group purchases, or whole herds. These may cover periods of specified intervals or an inventory period of one year. For any of these calculations a beginning and an ending weight are necessary. The beginning weight might be the purchase weight or the weight at the beginning of any check period including the beginning inventory. The ending weight might be the sales weight or the weight at the end of any check period including the closing inventory. Form 6.5 illustrates how the weight increase (total physical production) can be calculated for an inventory period of one year for beef and swine. The "produced" row is normally used only to show the number of animals weaned. This assumes that the "deaths" row shows only the number that died after weaning; however, if a separate record is kept for the breeding herd to include progeny until weaning and a separate record is kept for the feeding herd, this line could be used to show the weight and value of animals that were being transferred from the breeding herd to the feeding herd. This same value would be entered as sales for the breeding herd.

Form 6.4. Breeding Record

	Name or number	Date bred	Date due	Sire	Date born	No. born	No. weaned	Weaning weight	Comments
1	(Swine)								
2	54	Oct 28- Dec 12	Feb 19- Apr 5	Big Ham Feb 20- Long Boom Apr 1		450	4/5	—	
3									
4	50	Apr 16- May 16	Aug 8- Sept 7	Long Boom Aug 5- Steel gives Sept 10		300	280		Septo dinner
5									
6									
7									

Form 6.5. **Livestock Production Summary Record**

	Description	Beef No.	Weight	Amount	Swine No.	Weight	Amount	No.	Weight	Amount
1	Opening inventory	101	44,945	$10,357 00	469	79,910	$12,168 00			$
2	Purchases	80	55,890	12,435 00	3	805	445 00			
3	Produced				695					
4										
5	Total In	181	100,835	$22,772 00	1167	80,715	$12,613 00			$
6	Closing inventory	79	64,000	14,080 00	347	56,620	9,200 00			
7	Sales	100	95,331	23,130 00	804	193,356	30,809 00			
8	Home consumed products	1	750	160 00	4	950	160 00			
9	Deaths	1	800	—	10	1,495	—			
10										
11										
12	Total Out	181	160,881	$37,370 00	1165	252,421	$40,169 00			$
13	Inventory Change (6-1)	22	19,055	$ 3743 00	122	12,934	$ 2968 00			$
14	Net Increase (12-5)	xxx	60,046	$14,598 00	xxx	171,706	$27,556 00			$

Total Aggregate Livestock Increase $ 42,154 00

Form 6.5 uses the following formula to compute the amount of production that has taken place:

$$\left.\begin{array}{l} \text{Beginning inventory} \\ \text{Purchases} \\ \text{Production} \end{array}\right\} = \left\{\begin{array}{l} \text{closing inventory} \\ \text{sales} \\ \text{farm consumption} \end{array}\right.$$

The left-hand side relates to the total supply, and the right-hand side accounts for it. If all the items except production are available, that can be calculated be adding the closing inventory, sales, and consumption and subtracting the beginning inventory and purchases. This formula is useful for other tabulations such as determining the amount of feed fed and balancing livestock numbers. Physical quantities or dollar amounts work equally well.

Whether the weight and value of animals lost by death should be included when calculating increases is a debatable issue. It seems logical that from a feed efficiency standpoint, animals may have gained weight until death the same as those that lived. The farmer, however, never realizes the benefit from that increase. In Form 6.5 the weights of animals that died are included, but not the values. Note the balance of livestock numbers in Form 6.5. If these do not balance as in the case of swine, the farmer needs to begin looking for errors in his records.

Livestock feed records are among the most important statistical records if livestock production efficiency is to be measured. Feed costs account for 50-80% of the total cost of producing meat or other livestock products.

Feed records can be kept on individual animals, lots, pens, and purchases or for a whole class. Feed may be recorded daily or at intervals such as each week. Extrapolations can be made among measurement periods if records are not kept on a daily basis. Quantities can be measured by weight and volume or a combination of these. Form 6.6 shows one type of form that may be useful for cattle feeders.

Keeping feed records is tedious, but shortcuts may make the task easier:

1. The feed purchase record can be used to determine part or all the total feeds fed. If there is more than one livestock class and it is desirable to have separate feed records, the purchases should be separated by class. Also, if the types of feed are to be distinguished, these must be separated or at least identified in the purchase account so that separate tabulations can be made later. In all cases the quantity purchased should be recorded as well as the dollar amount. Adjustments for quantities on hand at the beginning and ending of the feeding period must be made. If the purchase account does not list all feeds fed (i.e., farm-produced feed), the feed purchase record must be supplemented with other feed information.
2. Feed processing records may be used to determine part or all of the feeds fed. If the quantity processed for a particular livestock class is recorded, this may add information to the purchase account or could constitute the total record. This information is readily available if the feed is processed by a custom operator or by a feed processing plant. For example, if all the shelled corn is fed to the swine, the number of bushels of shelled or cracked corn may be a good source of information for estimating the amount of corn fed to the swine.
3. The inventory method may be very useful to determine part of the total feeds fed. If a particular class of livestock is being fed out of one bin or granary, the quantity fed

Form 6.6. Livestock Feeding Record for _Heifers_

Day, week, or month	No. animals	Pasture Field	Pasture Days	Field	Days	Shelled corn (cwt.)	Ground ear corn (cwt.)	Oats (bu.)	Straw (bales)	Hay (bales)	Suppl. (cwt.)
1											
2											
3 Sept. 30, 1967	101	49	10					3		10	
4											
5 Sept. Total			10					30		100	
6 Oct 1, 1967	101	49	10								
7 Oct 10, 1967	101	456	21								
8											
9											
10 Oct. Total			31								
23 July 16, 1967	100					21					2
24						22					2
25 July Total						621					62
26						18					1.6
27						13					1.5
28						5					1
29											
30 Aug. Total						252					23
31 Annual Total						1208	1874	703	150	2637	283
						2157 lbs.	2346 lbs.			81 tons	

Form 6.7. Crop and Feed Balance

This table assumes the accounting of the operator's share only. Hence the feed purchased, raised, sold etc. refers to that belonging to the operator. Under livestock share arrangements it may be desirable to alter these measures to include the landlord's share.

	Unit of measure	Crop or feed							
		Corn		Oats		Hay		Swine supplement	
		Qty. (bu.)	Amount	Qty. (bu.)	Amount	Qty. (tons)	Amount	Qty. (cwt.)	Amount
1	Source Opening inventory	13,100	$13,1500.00	2200	$1430.00	60	$1080.00	95	$459.00
2	of Purchased	784	902.00	729	474.00			1056	6547.00
3	Feed								
4	Crops raised	17,122	17,978.00	3192	2075.00	82	1476.00		
5	Total In	31,006	32,060.00	6121	3979.00	142	2556.00	1141	7006.00
6	Closing inventory	15,520	16,296.00	2500	1625.00	56	1008.00	72	396.00
7	Sales			310	202.00				
8	Feed Landlord's share	1,824	1,952.00	729	474.00				
9	Accounted								
10	For								
11	Seeded			135	88.00				
12	Wastage, spoilage								
13	Total Out	17,344		3674	2389.00	56	1008.00	72	396.00
14	Total Feed Fed (5-13)	13,661	$13,612.00	2447	$1590.00	86	$1548.00	1069	$6610.00
15	Feed Swine	10,545	11,072.00	215	140.00			1069	6610.00
16	Consumed Beef feeders	2,950	3,098.00	2210	1436.00	85	1547.00	293	1448.00
17	by								
18	Livestock								
19	(from								
20	feed								
21	records)								

		Corn		Oats		Hay		Swine supplement	
		Qty. (bu.)	Amount	Qty. (bu.)	Amount	Qty. (tons)	Amount	Qty. (cwt.)	Amount
31									
32	Total Feed Recorded	13,495	$14,170.00	2425	$1576.00	85	$1547.00	1362	$8058.00
	Error (14-32)	166	$ -358.00	22	$24.00	1	$1.00	0	$.00

Form 6.8. Monthly Labor Summary

Month	Beef feeders Oper.	Family	Hired	Total	Hogs Oper.	Family	Hired	Total	Crops Oper.	Family	Hired	Total
					Enterprise							
Jan.	57		28	85	78		86	164	15		9	24
Feb.	53		23	76	76		146	222	16		6	22
Mar.	57		31	88	82		125	207	37		36	73
Apr.	57		16	73	58		114	172	86		80	166
May	65		26	91	43		89	132	86		98	184
June	44	15	22	81	31	15	81	127	61	72	83	216
July	28	24	15	67	41	49	51	141	68	85	66	219
Aug.	20	4	14	38	96	84	62	242	57	49	53	159
Sept.	39		20	59	45		93	138	40		58	98
Oct.	40		32	72	57		92	149	82		70	152
Nov.	41		31	72	54		87	141	66		69	135
Dec.	55	2	25	82	61	20	63	144	16		4	20
Total	556	45	283	884	722	168	1089	1979	630	206	632	1468

Form 6.8. (continued)

	Farm overhead				Total farm				Nonfarm			
Month	Oper.	Family	Hired	Total	Oper.	Family	Hired	Total	Oper.	Family	Hired	Total
Jan.	35		12	47	185		135	320				
Feb.	35		5	40	180		180	360				
Mar.	24		8	32	200		200	400				
Apr.	19		10	29	220		220	440				
May	26		7	33	220		220	440				
June	44	48	34	126	180	150	220	550				
July	38	17	18	73	175	175	150	500				
Aug.	27	13	21	61	200	150	150	500				
Sept.	76		29	105	200		200	400				
Oct.	21		6	27	200		200	400				
Nov.	39		13	52	200		200	400				
Dec.	48	18	8	74	180	40	100	320				
Total												

Note: Data are hypothetical but based upon farmer experiences.

121

could be obtained by taking beginning and ending inventories of feed in the bin.

4. The simplest method for calculating quantities of feed fed is to combine information from the purchases and sales accounts and the feed crop production record after adjusting for inventory change. The total quantity of feed available on the farm for all purposes is calculated by adding the beginning inventory, the feed produced, and the feed purchased. Subtracting the amount in the closing inventory and the amount sold from this total quantity leaves the balance used on the farm. Except for seed or other nonfeed uses this balance gives the total feed fed. This amount can be in dollars or physical quantities. Form 6.7 illustrates this method.

The bottom part of Form 6.7 shows how a feed record check can be made between individual feed records and the aggregate determinations described above. The major difficulty with this method is that it is aggregative across different types of feed and over livestock classes. Thus it is more useful for determining the value of feed fed than for the quantity. Where one livestock enterprise dominates the farm, the figures obtained are more meaningful. If there is only one major livestock enterprise, it may even be possible to separate the estimated quantity of feed fed to other minor enterprises, thus leaving a reliable estimate of feed fed to the major enterprise.

When making feed balance checks, some error is expected. Only when the error is large should the operator be concerned. Where errors exist, it is then a matter of judgment as to which record is the most accurate.

LABOR RECORDS

Records of labor use are helpful where there are particular labor problems. It is doubtful that most farmers would want to keep detailed records of labor use for all enterprises or activities. Such records may be justified for efficiency studies, for determining the total cost of an enterprise, or where labor allocation is important in the selection of enterprises. Labor becomes a particularly critical item at certain times of the year. As enterprises bid for labor use, its marginal value (measured in terms of increased net farm income) may rise many times over its market price. Thus farm labor records may be useful at intervals in the history of a farm even though they cannot be justified on a continuing basis by most operators. What is said for labor records can be applied also to farm machinery.

Alternatives to keeping detailed labor records can be found in agricultural extension service publications that report labor utilization for different enterprises and activities. These data are useful for planning but undoubtedly cannot be applied without adjustment to particular farming situations.

Form 6.8 illustrates a labor summary record for the Dunn farm. Furnishing information to this record is a labor log that reports the daily allocation of labor for each worker.

A complete labor record provides summaries of the following information:

1. Total labor used by each enterprise
2. Seasonal use of labor by enterprise or activity
3. Time required to perform particular activities, i.e., plow, feed cattle, milk cows, etc.

These data are all useful for farm planning at various stages. They are essential when determining the substitution of machines and equipment for labor.

7

ANALYSIS OF THE TOTAL FARM BUSINESS

FARM ANALYSIS IS THE CLIMAX of any record-keeping activity. It is only through this phase that the farmer gains insights into his business. Without it the record-keeping function is little more than accounting for tax liability. This has a negative ring notwithstanding the need of a good set of records for this purpose. Up to this point record keeping has been a cost both in time and aggravation. Accurate records demand consistency and attention to details which at times become drudgery. The paycheck at the end of the work period keeps most laborers on the job. This is no less true for farmers who are concerned with accounts and records. The size of the paycheck depends upon the use that is made of the completed product—the farm accounts and records.

It is important that analysis includes the same set of measurements and elements from year to year so that meaningful comparisons can be made over time. Due to the variable nature of farm prices and production, any one year's results may not be too indicative. This, however, does not limit the scope of the analyses that may be desirable in any one year.

The continuing analyses made from year to year should be similar in nature and composition to the farm record summaries published by cooperative extension services or farm record-keeping associations in your area. Comparison of production is an enjoyable activity for many farmers and in this case may serve to give many helpful leads for business improvement. In Iowa, summaries from Iowa farm business associations are published annually for every section of the state. Similar summaries are available in other states. These provide much useful data for individual farm comparison and evaluation.

The discussion that follows will not completely cover the many analyses that could be made but will concentrate upon measurements meaningful to the majority of farmers; i.e., the ones usually found in the summaries of farm record-keeping associations. A little detail in the inventory and in receipt and expense accounts can be very rewarding to the farmer; however, elaborate sets of records are not necessary to make meaningful and broad analyses.

Not all the records discussed thus far are required for making analyses related to the total farm. If enterprise analyses are desired, some records in addition to those discussed may be necessary, but it is surprising how much information can be gleaned from a regular set of farm accounts and records about particular enterprises. Even though the data used in the analyses may not be 100% accurate in all cases, they may be good enough to guide the business out of failure and into success.

DISTRIBUTION OF NET FARM INCOME

Net farm income was previously defined to be:

1. A return to unpaid (net worth) capital
2. A return to unpaid labor
3. A return to management

If opportunity returns are imputed to any two of these, the remaining residual of net income can be said to be a return to that remaining factor. There is no easy and accurate way of determining the contribution of working and fixed capital or operator labor and management to net farm income. Whatever allocation is made will be arbitrary. However, to obtain some relative measure of efficiency in the use of capital, labor, and management, this allocation process is useful. For labor and capital there is a well-established market price that can be imputed as a cost to the business for its use. (This also could be viewed as a return to these two input factors.) This is not true of management. From an owner-operator standpoint the management return is similar to profit (also defined as rent by some economic texts), which is the residual left after all other factors have been paid a market price.

Table 7.1 illustrates the distribution of net farm income to labor, management, and capital. The distributions to management and capital are the most meaningful. Note first that this allocation process can be just for the operator or the landlord or may be combined for the total farm. The "total farm" column represents the situation where the operator is the sole owner. If comparisons are to be made with other farms, the total farm approach is the only one that has comparable meaning.

The total asset values are repeated in Table 7.1 for easy reference. These are average values for the year and are determined by adding the beginning and ending inventories and dividing by two. Some accounting systems use only the ending inventory values. This is permissible if applied consistently, but it gives a less representative estimate of the business assets used during any one accounting period.

In order to have measures that can be compared meaningfully between farms, it may be necessary to adjust the net farm income since not all farms have the same amount of liabilities (hence interest) and family labor. The most reasonable base for comparison is the farm that is wholly owned by the operator with no family labor furnished. Thus an adjusted net farm income is calculated by adding to the net farm income the interest that had previously been counted as a business expense and subtracting from net farm income the value of family labor that has not received a farm wage payment. Recall from the discussion on interest payments in the section on income and

TABLE 7.1.   Distribution of net farm income for the Dunn farm, 1972

Item	Operator	Landlord	Total farm
Total asset value (from net worth statement):			
Operating capital	$ 43,926	. . .	$ 43,926
Working capital	16,165	. . .	16,165
Fixed capital	112,952	$62,000	174,952
	$173,043	$62,000	$235,043
Adjusted net farm income:			
Net farm income	$ 14,703	$ 3,583	$ 18,286
Plus interest expense	3,703	. . .	3,703
Minus value of unpaid family labor	−600	. . .	−600
	$ 17,806	$ 3,583	$ 21,389
Return to operator labor and management:			
Adjusted net farm income	$ 17,806	$ 3,583	$ 21,389
Minus 5% interest return on total asset value	−8,652	−3,100	−11,752
	$ 9,154	$ 483	$ 9,637
Return to management:			
Return to operator labor and management	$ 9,154	$ 483	$ 9,637
Minus value of operator's labor	−6,000	. . .	−6,000
	$ 3,154	$ 483	$ 3,637
Return to capital investment:			
Adjusted net farm income	$ 17,806	$ 3,583	$ 21,389
Minus value of operator's labor and management	−6,000	. . .	−6,000
	$ 11,806	$ 3,583	$ 15,389
Percent return to capital			
(return to capital ÷ total asset value):	6.8%	5.8%	6.5%

expense that interest can be either an expense or a return. The adjusted net farm income as now calculated and shown in Table 7.1 is defined to mean:

1. A return to asset capital
2. A return to operator labor
3. A return to management

The return to management is calculated by subtracting from the adjusted net farm income an opportunity return (a market price) for the use of the operator's labor and total asset capital. A return to the operator's labor and management is first calculated by subtracting a return for the asset capital. In the example in Table 7.1 a 5% return is applied to the total capital asset value. It may be more desirable to use a varied return rate for each of the three types of capital, e.g., operating capital 6%, working capital 5%, and fixed capital 4.5%. The rates used should be in line with competing uses for this capital. The farmer needs to ask himself what he could expect to receive if he loaned his money or invested it alternatively, e.g., in government bonds or corporate stock. The straight 5% is used here because of its simplicity for illustration. The return to operator labor is based upon farm labor wage rates and alternative nonfarm employment opportunities.

The return to capital is calculated in much the same way except that the opportunity return for the operator's labor and management is subtracted from the adjusted net farm income. The reader may be surprised that the same figure, $6,000, is used to represent both the value of the operator's labor and his labor and management. This says that his management value is zero. This may or may not be true depending upon the operator's management skills in alternative employment. It is common practice in the allocation process to use the same figure because it is more uniform for between-farm comparisons and the opportunity value of the operator's management is not known.

Since the return to capital is generally expressed on a per dollar basis in percentages, it is necessary to divide the residual return to capital by the total asset capital figure. Percentages thus obtained are then comparable to interest rates and returns in alternative investments and among farms, regardless of size.

One useful calculation not illustrated in the example is the percent of return to operator's net worth capital. Many investors make their living by expanding their limited capital through credit in order to make relatively large investments possible. If the return on the total investment is only slightly larger than the interest cost, a large return can be realized on a limited investment. For example, assume a $10,000 net worth that is used to finance a $100,000 investment ($90,000 is obtained through credit). If the interest cost is 6% and the net return to total capital is 7%, the return to net worth capital would be 16%. Interest cost $90,000 x .06 = $5,400; return to total capital $100,000 x .07 = $7,000; return to net worth capital $7,000 — $5,400 = $1,600; percent return to net worth capital ($1,600 ÷ $10,000) x 100 = 16%. On the Dunn farm the return to net worth capital is calculated as follows (see Tables 5.3, 4.5):

1. Net farm income	$14,703
Minus: Value of unpaid family labor	600
Minus: Value of operator's labor and management	6,000
2. Return to net worth capital	$8,103
3. Percent return to net worth capital	

$$\$8,103 \div \left(\frac{\$103,146 + \$106,963}{2}\right) \times 100 = 7.7\%$$

If, over a period of years, the farmer cannot experience an opportunity return for the use of his resources (net worth capital) and his labor and management, perhaps he should consider other employment. On the other hand, if his returns are high, perhaps he should consider expanding his business. Regardless of the size of the calculation it is indicative of the efficiency with which the farm resources are utilized, gives an indication of problem areas and successes, and lends direction to the business.

SOURCES OF INCOME

Gross profit is defined and discussed in Chapter 5 in connection with the income statement. It is defined as a measure of gross value increase from productive farm activities and is calculated by adding the sales from crops, livestock, livestock products, and other miscellaneous income; adjusting for inventory changes of crops and livestock; and subtracting livestock and feed purchases. The sources of or contributors to gross profit can be found by looking at the farm enterprises that usually show up in the receipt section of the account book. Since the total farm is being analyzed, attention is centered upon total farm, both operator's and landlord's shares. For the Dunn farm gross profit was $44,858 (Table 5.5).

Since gross profit measures the value of production increases on the farm, it should be possible to reconstruct this figure more directly from the value of crops and livestock production. The total value of crop production for the Dunn farm was calculated to be $27,471 (see Form 6.2). This measured the gross value (also gross profit from crops) of all crops produced, both operator's and landlord's shares. It is optional whether diverted acreage payments and other similar crop payments should be included in the value of crops produced or should be shown as a separate source of income. The $27,471 included government program payments.

Gross profit from livestock production is calculated as the increase in livestock value over the total feed cost. The value increase is determined by subtracting purchases from sales and adjusting for inventory changes. For the Dunn farm this figure was $42,154 (see Form 6.5). Feed costs include purchased feed and the value of farm-raised feed fed. This figure can be calculated from feed records or estimated from crop production records, purchase and sales accounts, and crop inventories. For the Dunn farm the total value of feed fed was $25,379. This figure is tabulated in Table 7.2. This tabulation for individual feeds fed by livestock class was illustrated in Form 6.7. The total feed fed is a summation of the amount of the individual feeds. Subtracting the value of feed fed, $25,379, from the value of livestock increase, $42,154, gives a gross profit from livestock production of $16,775.

Other sources of gross profit may come from custom work off the farm, dividends from farm cooperatives, gasoline tax refunds, etc. These generally are included as miscellaneous (or other) income or miscellaneous gross profit. For the Dunn farm this figure was $370.

Adding together the gross profits from crops, livestock, and miscellaneous sources gives total gross profit for the farm of $44,616 ($27,471 + $16,775 + $370).

This gross profit figure is $242 less ($44,858 − $44,616) than the one calculated in the income statement (Table 5.5). This is accounted for largely by the value of crops kept for seed, which did not show up on the income statement but is considered when calculating the value of crops raised. Also, prices used for inventories and crop values may not correspond to sales prices. For these reasons it is not expected that the gross profit figures calculated from the two different methods will be the same; but they should be nearly the same, give or take one or two hundred dollars. Some farm accountants calculate gross profit from miscellaneous sources as the difference between the total gross profit figure calculated in the income statement and the sum of gross profits for the major enterprises. For the Dunn farm this figure would be $612 ($44,858 − $27,471 − $16,775) instead of $370. Gross profit for the whole farm would remain at $44,858.

TABLE 7.2. Value of feed fed on the Dunn farm, 1972

Add		Subtract	
Beginning feed inventory	$16,519	Closing feed inventory	$19,641
Value of crops raised	27,471	Feed crop sales	202
Feed purchased	9,371	Value of crops used for seed	88
		Value of soybeans raised	2,870
		Government program payments	2,755
		Crop-share rent	2,426
	$53,361		$27,982
Value of Feed Fed	$25,379		

The difference between gross profit and net farm income is farm expense. For the Dunn farm, gross profit of $44,858 minus net farm income of $21,989 gives a business expense of $22,869. The $21,989 net farm income used did not include interest as an expense. The concept is still that of the total farm as viewed from the asset side of the net worth statement.

From this analysis of gross profit some ideas can be gleaned about the important sources of or contributors to the business income. Gross profit will be used many other places in the analysis and its importance will be discussed further.

INCOME RATIOS

Rate of capital turnover is a common measure of business efficiency. The profit margins of many companies are very low per unit of production; but where the volume of business (turnover) is large, the returns to capital and management could be also. Meat packing plants are a case in point where the margin per pound of meat sold is less than 1% in many cases. Rate of turnover is no less important to the farmer.

Rate of capital turnover can be calculated in two ways with each carrying a little different meaning. One method is to divide gross income by total asset capital; the other is to divide gross profit by total asset capital. Using gross profit for the Dunn farm, the capital turnover rate is calculated as follows:

(Gross profit ÷ capital managed) x 100
($44,858 ÷ $235,043) x 100 = 19%

Using gross income this calculation would be:

(Gross income ÷ capital managed) x 100
($67,109 ÷ $235,043) x 100 = 29%

The method chosen will depend upon the comparisons being selected and their use.

The rate of capital turnover in most farming activities is low when compared to many other businesses. The 29% for the Dunn farm illustrates this. In comparison, for 1967 a grain marketing cooperative in Iowa showed over 500% turnover, a regional farm sales cooperative over 300%, and a national meat packing plant over 700%.

Within the farming industry the rate of turnover varies with the enterprises selected and the capital position of the operator. The turnover rate is higher for livestock than for crops, higher for dairy cows than beef cows, higher for broilers than layers, etc. Consider beef cows and feeder cattle as an example. Using some average figures for the Midwest, turnover rates are determined as shown in Table 7.3. Farmers short on capital may wish to emphasize enterprises that have a relatively high turnover rate, providing there is also an adequate profit potential. This aspect is particularily important to beginning farmers.

Farmers that rent have a higher turnover rate than owner-operators. Renting is a means of spreading limited net worth capital over a larger total capital base of operation. The rate of turnover on net worth capital is increased considerably by this means. For the Dunn farm this would be as follows:

(Operator gross profit ÷ net worth) x 100
($42,095 ÷ 105,054) x 100 = 40%

TABLE 7.3. Determination of rate of turnover

Item	Returns	Investment	Percent turnover
Using gross income:			
Beef cows	$125	$380	33
Beef feeders	250	202	111
Using gross profit:			
Beef cows	49	380	13
Beef feeders	26	202	13

This compares to 19% for the total farm as previously calculated.

Gross profit per dollar of expense also reveals income efficiency. Net income by itself is one of the best overall efficiency measures. This ratio relates to the income-producing capacity of expenses. It is calculated by dividing gross profit ($44,634) by business expenses ($22,869). For the Dunn farm this calculation is as follows:

$$\text{Gross profit} \div \text{expenses}$$
$$\$44,858 \div \$22,869 = \$1.96$$

It cannot be interpreted that if another $1.00 is expended $1.96 more of production value will be created. Gross profit per dollar of expense is an average concept and not a marginal one. However, it does relate to the efficiency with which resources are employed and products are marketed.

Other meaningful income relations can be calculated such as net income per rotated acre or per hour of operator labor. For the Dunn farm these two measures are calculated as follows:

Net income per rotated acre:
$$\text{Adjusted net income} \div \text{rotated acres}$$
$$\$21,389 \div 343 = \$65.36$$

Net income per hour of operator labor:
$$\text{Adjusted net income} \div \text{total operator labor}$$
$$\$21,389 \div 2,640 = \$8.10$$

CAPITAL RATIOS AND POSITION

Asset to liability ratios were presented earlier in the net worth statement discussion (see Table 4.5). The net capital ratio, working capital ratio, and current capital ratio were treated. That discussion has relevance here as part of this discussion on capital efficiency.

Capital per $100 gross profit or gross income, capital per $100 net income, and capital per man are important capital efficiency ratios. For the Dunn farm these are calculated as follows:

Capital per $100 gross profit:

$$(\text{Total capital} \div \text{gross profit}) \times 100$$
$$(\$235,043 \div \$44,858) \times 100 = \$524$$

Capital per $100 net income:

$$\text{(Total capital} \div \text{adjusted net income)} \times 100$$
$$(\$235,043 \div \$21,389) \times 100 = \$1,099$$

Capital per man:

$$\text{Total capital} \div \text{man-years}$$
$$\$235,043 \div (28 \text{ months} \div 12) = \$100,747$$

These latter capital ratios are inverses of the income ratios already discussed and have some of the same meaning.

FARM SIZE EFFICIENCY

Economies of size are reflected in most other measures of efficiency.

Size relates to the idea that it is necessary to have a large volume of business in order to have a large net income. However, there are diseconomies of size as well as economies. A farmer must also be efficient in the use of his resources and the spending of his money. Thus size is not a sufficient condition for large net income.

Farm size can be measured in a number of ways, just as that of a man. Some men are judged on the basis of height, some by weight, some by their brawn, and others by the size of their capital investments or income. These do not all measure the same dimension of size and must be interpreted for each individual or purpose. Farm size may be measured by the size of the capital assets; inputs of land, livestock, or labor; or volume of income produced.

Size efficiencies for Iowa farms in 1971 are shown in Table 7.4. Several things can be observed by studying this table. It is interesting that expenses per acre were reduced by $30 per acre between the 240- and 600-acre farm when family labor was included as an expense. Putting it another way, one 600-acre farm has $18,000 less total expense than two and one-half 240-acre farms. The gross profit item also is interesting in that as farm size becomes larger gross profit per man from livestock did not increase nearly so much as for crops. This latter relationship may not hold true for other farming areas or other years. Also note that man-months of labor was not even doubled as farm size increased by 3.7 times between 160 and 600 acres.

For the Dunn farm the following quantities are calculated:

1. Acres:
   Total operated = 400
   Crop (84% of total) = 337
   Rotated (86% of total) = 343
2. Total capital managed (year average):
   ($229,177 + $240,909) ÷ 2 = $235,043
3. Gross profit = $44,858
4. Man-months of labor: 28
   Operator = 11
   Hired = 12
   Family = 5

The *acre* is the most common unit for determining farm size. This is a satisfactory

TABLE 7.4.   Size efficiency measures as related to income and expenses on Iowa farms, 1971

Measures	Size of farm (acres)				
	160	240	320	440	600
Gross profit per man:					
Livestock production	$ 9,054	$ 9,256	$ 9,699	$10,626	$12,015
Crop production	11,183	16,162	21,170	24,818	32,496
Miscellaneous	1,914	1,071	1,343	1,559	1,661
Total	$22,151	$26,489	$32,212	$37,003	$46,172
Expenses per acre:					
Machinery costs	$ 41.69	$ 31.79	$ 28.30	$ 25.20	$ 22.27
Taxes, insurance, buildings	22.60	17.80	16.28	15.27	13.71
Crop expenses	20.03	17.55	17.15	17.23	18.87
Hired labor	3.59	3.39	3.49	3.90	5.70
Other	8.81	5,89	4.55	3.74	3.21
Subtotal	$ 96.72	$ 76.42	$ 69.77	$ 65.34	$ 63.76
Operator and family labor	41.56	27.13	19.53	15.49	9.83
Total	$ 138.28	$ 103.55	$ 89.30	$ 80.83	$ 73.59
Man-Months of Labor	13.6	14.3	15.3	16.8	23.1
Machinery Investment per Acre	$ 84	$ 63	$ 57	$ 52	$ 50

Source: 1971 Cost and Returns on Iowa Farms, Iowa Coop. Ext. Serv. FM 1640, Ames, 1972.

measure where the major source of income is from crop production and similar crops are being produced. For Iowa where 65-75% of gross profit is from crop production and the major cash crops are corn and soybeans, the acre is a good unit of measure. It is not a good unit of measure where land productivities differ greatly or the major enterprises are not the same. For some comparisons the number of acres in rotation may be a better measure than total acres.

Capital managed is a more broadly oriented measure of size. It takes into account land productivity differences as reflected in land values and is aggregative of different enterprise investments. However, it is less well understood and not easily derived. Also, it may not accurately describe size as it relates to income where the turnover of capital in the productive activities differs greatly.

Gross profit (or income) measures size in relation to the income generating power of the farm. It, like capital managed, can be broadly applied but is probably less well understood than either of the other two measures; however, it may more accurately describe size than either. It would not be a good measure of size where business expenses, as related to production activities, used up greatly different percentages of the gross profit. Vegetable production with very high expenses compared to dryland wheat production illustrates this.

Man-months of labor as a measure of size relates to one of the major input factors in many production situations. At a time when labor accounted for a large portion of business expenses or inputs it was a better unit of measure than when capital is used to replace much of the labor input. However, where enterprises are similar and production practices uniform, months of labor are still a good unit for measuring size. As can be seen in Table 7.4, the man-months of labor used do not increase proportionately to acres.

Units of livestock may be one of the best measures of size in many areas where livestock contributes the major portion of income. The number of livestock breeding units, the number of feeders, the number of layers, or the number of poults may be the most descriptive and meaningful unit for describing the size of business in many areas of the country. A 400-cow unit is very descriptive of size to the cattle rancher in Wyoming, Nevada, or New Mexico.

Size is an important aspect in all expansion considerations on the farm. Whether or not an expansion or increase in size will reduce unit costs and increase income must be determined by budgeting or programming each individual situation. There is a different cost curve for each farm. A study of the farm records should give the farmer some indication of his position on the cost curve and provide some directional guides for decision making.

CROP EFFICIENCY

Measures of income, investments, and production could be generated for each major enterprise on the farm. (Enterprise accounts will be covered in a later section.) The purpose here is to look at general measures as related to the whole cropping program. The importance of crops in generating gross profit and the percentage of total acres in rotation are both important concepts that have been discussed earlier but will not be further elaborated here.

Yield is the commonest and perhaps the most useful measure of crop efficiency. This is particularly true where crops contribute the bulk of gross profit. It is a measure that farmers know and understand well. Crop yields are a common topic of discussion in many farmer meetings. Comparisons with neighbors, record-keeping groups, and county or township reports are easily made. Discussions and comparisons involving varieties, fertilizers, cultural practices, pests, etc., as related to yield can be very beneficial, particularly to the less efficient producer.

It should be realized that yield can be too high as well as too low. Economic considerations will dictate the most profitable level of production for each farm.

Yields for the Dunn farm are summarized in Table 7.5. A *crop yield index* that compares composite crop yields for a farm with some norm may be a useful figure. This determination for the Dunn farm is illustrated in Table 7.6. The method shown in Table 7.6 for calculating a crop yield index weighs the individual crop yield differences by the number of acres for each crop. This is accomplished by determining the number of acres required at community yield levels (288) (or some other norm) to produce the amount of crops on the reference farm. Dividing these acres by the number of acres on the reference farm (294 acres) and multiplying by 100 gives a percentage of comparison (98%).

The *distribution of crops* may also be important in relation to net farm income. Not all crops have the same income-producing potential. Table 7.7 shows some average income figures for major crops in Iowa.

It should not be interpreted that individual farmers will always achieve the average. For example, some farmers may find oat production more profitable than soybeans or

TABLE 7.5. Summary of yields for the Dunn farm, 1972

Crop	Acres	Percent of rotated acres	Yield
Corn (bu)	173	50	99.0
Oats (bu)	61	18	52.3
Soybeans (bu)	37	8	32.3
Hay (T)	23	7	3.5
	294		

TABLE 7.6. Determination of crop yield index

Crop	Production (Dunn farm)	Acres at Community yields*	Acres required for reference farm at community yields
Corn	17,122 bu	102.5	167
Oats	3,192 bu	57.0	56
Soybeans	1,196 bu	27.0	44
Hay	82 T	3.9	21
			288

Crop Yield Index = (288 acres ÷ 294 acres)100 = 98

*Hypothetical figures.

corn on parts of their farm. Also, some unmeasured costs and values are not included here, such as straw from oats and nitrogen from soybeans and alfalfa. These figures emphasize that the combination of crops selected can be an important consideration in maximizing net farm income. These percentages are listed for the Dunn farm in the discussion on crop yields.

Value of crops per rotated acre relates to the distribution of crops just discussed. This figure is calculated by dividing the gross profit from crops by the number of acres in rotation. For the Dunn farm this is calculated as:

Gross value of crops produced per rotated acre = $27,471 ÷ 343 = $80

Fertilizer and lime cost per rotated acre is another important efficiency measure since fertilizer has become such an important crop input. For the Dunn farm this is calculated as:

Fertilizer and lime costs per crop acre = $3,963 ÷ 337 = $12

Crop efficiency measures and their importance may be summarized as follows:

1. Farms with high crop value per tillable acre have higher net incomes than those with low crop values within the same size group.
2. High crop yields contribute to high crop values per acre.
3. Crop expenses (largely seed and fertilizer) average higher for farms with high crop values per acre and these farms still have higher net incomes.

TABLE 7.7. Profitability of major Iowa crops (1970-72 average)

Crop	Yield/acre	Price	Income per acre	Operating cost per acre	Income over cost per acre
Corn	101.3 bu	$ 1.21/bu	$122.57	$56.88	$65.69
Soybeans	33.7 bu	3.44/bu	115.93	33.89	82.04
Oats	56.7 bu	0.73/bu	59.39	29.10	30.29
Alfalfa hay	3.15 T	21.80/T	68.67	43.76	24.91

Note: Yields and prices are based upon state averages for the period. Costs were derived from averages developed from farm costs and returns studies. Labor and land costs were not included.

Source: Present and Future Changes in Iowa Agriculture, Iowa Coop. Ext. Serv., FM 1622 (rev.), Ames, 1973.

LIVESTOCK EFFICIENCY

Production and feed determinations are the major efficiency measures that can be developed from the general set of farm accounts and records for livestock. Measures of production relate to output per livestock unit such as pigs farrowed or weaned per litter, percent of calf crop or pounds of calf weaned per cow, milk per cow, eggs per hen, and weight gain per day. For the Dunn farm the following measures were calculated:

Hogs:
Number of litters farrowed = 104
Pigs weaned per litter = 695 pigs weaned ÷ 104 = 6.7
Feeder cattle:
Number of feeders sold = 100
Rate of gain (heifers): Weight gain ÷ feeder days = 45,525 ÷ 30,273 = 1.5 lb/day

For the feeder cattle "weight gain" was determined by purchase and sale records. "Feeder days" is a summation of the number of days each animal was on the farm. Feeder days can be tabulated easily by using purchase and sale dates. Extreme accuracy is not essential for many of these tabulations. For the most part each measure is only a clue to be used as evidence for detecting some suspected problem.

The importance of feed efficiency has already been discussed as well as tabulation procedures for all livestock and individual enterprises. Feed efficiency can be a physical or financial measure. Since tabulation aspects are similar, only the financial measure will be illustrated. However, keep in mind that the amount of feed required to produce a pound of meat, a dozen eggs, 100 pounds of milk, etc., can be determined by using the same procedures. In most cases the only requirement is to keep a record of the weights of feed, livestock, and products purchased and sold.

Returns per $100 feed fed is the commonest measure of feed efficiency. This can be calculated for a composite of all classes of livestock on a farm from very few additional records beyond those required for income tax purposes. Production and feed tabulations have been previously illustrated for the Dunn farm (see Forms 6.5 and 6.7). Returns per $100 feed fed are given below:

Livestock increase ÷ feed fed
($42,154 ÷ $25,379) x 100 = $166

Where there is only one major livestock enterprise the all-livestock returns per $100 feed fed figure can be very meaningful. It is less meaningful but still a useful figure for comparison if the livestock on the reference farm is typical of the area being compared. Some comparative returns per $100 feed fed for Iowa are shown in Table 7.8.

MACHINERY EFFICIENCY

Investment and cost per rotated acre are major efficiency measures that can be developed without rather detailed records of machine use. This detail can only be justified for individual machines where particular information is desired. The

TABLE 7.8. Livestock income per $100 feed fed for Iowa, 1965-74

Year	Corn Price	All Livestock	Hogs	Fed Cattle	Beef Cows	Dairy	Hens
1965	1.13	$167	$189	$154	$136	$160	$120
1966	1.19	162	198	123	153	168	158
1967	1.13	152	173	128	160	182	98
1968	1.04	166	180	151	171	199	120
1969	1.08	181	209	152	181	197	160
1970	1.17	159	186	119	184	200	148
1971	1.21	156	151	150	182	194	145
1972	1.11	198	218	172	203	206	119
1973	1.81	180	193	139	239	184	165
1974	2.87	113	140	63	131	145	117
10-yr. avg.	$1.37	$163	$183	$135	$175	$184	$135
High, 2-yr.	$2.34	$190	$214	$163	$221	$203	$162
Low, 2-yr.	$1.06	$132	$146	$ 91	$134	$152	$108

Source: Iowa Coop. Ext. Serv., Ames.

discussion here treats the total investment without reference to types of machine or use.

Investment per rotated acre is calculated by dividing the average of the beginning and ending inventories by the number of rotated acres. For the Dunn farm this is calculated as follows:

Investment per acre (ave.) = $12,297 ÷ 400 A = $31
Investment per rotated acre (ave.) = $12,297 ÷ 343 A = $36

Cost per rotated acre is calculated by adding operating and investment expenses for machinery and equipment. These figures can be found in the income statement. For the Dunn farm, machinery costs were calculated as shown in Table 7.9. Caution should be exercised in interpreting these figures. Figures that are too low as well as those that are too high may indicate machinery investment and cost problems. These figures must be interpreted with regard to farm size, crops and livestock produced, labor availability, and alternative returns on investment. As with other efficiency measures these figures are averages and do not indicate the return on additional investment. The division into fixed and variable costs may have some added benefits for farm planning. The variable costs will change in response to fluctuations in rotated acres.

TABLE 7.9. Machinery and equipment cost per rotated acre

Variable costs		Fixed costs	
Machinery and equipment hire	$1,682	Depreciation	$3,734
Machinery and equipment repair	1,464	Taxes	320
Fuel and lubricants	1,163	Insurance	60
Farm share of auto expense	117		$4,114
Farm share of utilities	232		
	$4,658	Total All Costs	$8,772
		$8,772 ÷ 343 rotated acres = $26/A	

LABOR EFFICIENCY

Even though labor may not be as important as it once was, it is still a major input on most farms. The input unit is usually measured in man-months or man-years of labor. Common measures relate to physical output or value. Typical of these tabulations are the measures illustrated below for the Dunn farm:

Labor efficiency:
Months of man-labor = 28
Rotated acres per man = 343 ÷ (28 ÷ 12) = 147
Value of crops produced per man = $27,471 ÷ (28 ÷ 12) = $11,790
Value of livestock increase per man = $42,154 ÷ (28 ÷ 12) = $18,069
Gross profit per man = $44,858 ÷ (28 ÷ 12) = $19,228

Productive man-work units was a common measure when labor contributed a larger share of farm inputs. It is still used in some states as a measure of farm size and labor use efficiency. One man-work unit measures the amount of work a typical man would accomplish under average farm conditions at usual productive farm tasks in a ten-hour day. Agricultural experiment stations have issued average man-work units for each farm activity such as production from one acre of corn, oats, or hay or from each head of dairy cows or feeder steers or from 100 chickens. By multiplying the number of units of each of these activities by their respective man-work unit standards, the total man-work units for the farm could be determined. The labor actually used on the farm could then be compared to measure the farm efficiency as related to the standard or average.

KEY ANALYSIS FACTORS

When making an analysis of the total farm business, the following five key factors are the most important and probably the minimum needed for successful or adequate analysis information. The factors are:

1. Gross value of crops per rotated acre
2. Livestock returns per $100 feed fed
3. Gross profit per man
4. Gross profit per $100 expense
5. Power and machine costs per rotated acre

An analysis of these five factors will often account for 75% or more of the income difference between farms when a comparison is made. Farms with above average performance in all five factors are high-profit farms a very high percentage of the time. Farms that are below average in all five factors (or most of them) are low-profit or low-income farms a high percentage of the time.

Gross value of crops per rotated acre indicates a combination of the selection of crops and yields realized. This provides some guidelines to the performance of the cropping program relative to the value of the land resource being used. Livestock returns per $100 feed fed indicate the margins realized on livestock over feed costs. This margin must pay for costs other than feed and provide any profit. The other costs over feed include interest, labor, facilities, supplies, veterinary and medical, etc. The return per $100 feed fed for different livestock varies depending upon the livestock and will

TABLE 7.10. Is Yours a High- or Low-Profit Farm?

Item	Average	High-profit farms	Low-profit farms	Your farm
Income and expenses:				
Income less feed and livestock purchases	$ 44,713	$ 63,370	$ 32,299	$ 44,858
Less expenses	26,025	30,904	25,111	22,869
Net farm income	$ 18,688	$ 32,466	$ 7,188	$ 21,989
Charges for operator and family labor, land and capital used	14,923	15,987	14,253	17,752
Management return	$ 3,765	$ 16,479	-7,065	$ 4,237
Resources used:				
Land (total acres)	289	294	289	400
Land (rotated acres)	255	266	252	343
Land (total value)	$125,948	$119,033	$126,011	$ 171,465
Machinery and equipment	$ 14,983	$ 17,674	$ 14,569	$ 12,297
Livestock and feed	$ 58,802	$ 86,070	$ 44,571	$ 43,257
Months of labor	15.5	16.4	15.3	28
Sources of gross profit:				
Livestock increase	$ 52,404	$ 84,499	$ 34,672	$ 42,154
Feed fed	34,547	51,866	27,270	25,361
Livestock increase over feed	$ 17,857	$ 32,633	$ 7,402	$ 16,793
Crop production	24,812	27,724	23,394	27,471
Miscellaneous	2,044	3,013	1,503	594
Gross profit	$ 44,713	$ 63,370	$ 32,299	$ 44,858
Efficiency of operations:				
Corn yield/acre (bu)	100	106	95	99
Crop value/rotated acre	$ 96.77	$ 105.56	$ 92.65	$ 80
Machine and power costs/rotated acre	$ 36.45	$ 47.24	$ 36.13	$ 26
Livestock returns/$100 feed	$ 153	$ 163	$ 127	$ 166
Rotated acres/man	197	184	190	147
Gross profit/man	$ 34,616	$ 44,690	$ 24,374	$ 19,228
Gross profit/$100 expense	$ 172	$ 204	$ 126	$ 195
Gross profit/$100 invested	$ 22	$ 28	$ 18	$ 19
500 Acre and over size group:				
Management return	$ 18,014	$ 42,921	$ -3,660	$ 3,637
Gross value of crops/rotated acre	97.72	99.13	92.22	80
Livestock returns/$100 feed fed	158	164	141	166
Machine and power cost/rotated acre	30.14	35.33	28.09	26
Gross profit/man	48,825	57,073	39,242	19,228
Gross profit/$100 expense	178	205	137	195

differ from year to year. Therefore, a comparison of similar farms for the same period provides good guideline data for determining the profitability of the livestock enterprise on an individual farm.

Comparison with other farms has been mentioned throughout the discussion on efficiency. A comparison of the Dunn farm with other farms in northwest Iowa is shown in Table 7.10.

Gross profit per man is the only efficiency factor that indicates the volume of production. It shows the volume being produced and implies or gives some indication of labor efficiency. Gross profit per $100 expense provides some guidelines on the overall efficiency of the business. It indicates whether margins over costs are satisfactory to produce profits. Machine and power costs per rotated acre indicate whether the farm may have excessive costs in this area. These costs must be related to volume of business, yields, and livestock production to be meaningful.

STANDARDS OF COMPARISON

There are three major levels of comparison for the farm analysis:

1. A comparison with the farm plan that was developed at the beginning of the accounting period
2. A comparison over time within the farm
3. A comparison with other farms

The end-of-the-year results measure the effectiveness of the farm plan or budget; a disappointing outcome does not mean the plan was poor. There are too many uncertainties that cannot be quantified when making the farm budget to expect the results to be exactly as planned. Aside from the uncertainties there are also misjudgments, incorrect information, etc., that can be improved in future planning activities by seeing what went wrong.

A comparison of efficiency determinations over time measures the movement of the farm toward the goals and objectives of the farmer. These might be a larger farm, more pigs weaned per litter, the highest corn yield in the county, or a larger net farm income. Regardless of the goal, if it can be quantified, records are a useful tool to measure its accomplishment.

8

ANALYSIS OF INDIVIDUAL ENTERPRISES

ALL THAT HAS BEEN SAID for the total farm in the previous chapters can be applied to individual enterprises. In fact, farmers who specialize in only one product such as swine production, cattle feeding, or corn production have only one enterprise account. However, this is not the usual case, particularly in the Midwest. Most midwestern farms have a minimum of two crop enterprises and one (sometimes two) livestock enterprise. Where there is more than one major enterprise, it may be desirable to make some analyses to determine efficiency or to obtain data for farm planning activities.

Much information relating to individual enterprises can be gleaned from the general set of farm accounts and records if only a small amount of detail is added. The purpose of this discussion is to indicate the additional detail needed, illustrate how an enterprise analysis can be developed, point out some of the more important information that can be obtained from an enterprise analysis, and highlight items the manager can use in planning.

A degree of caution should be exercised in determining the amount of detail to include in an enterprise analysis. It is important to concentrate on the data that are the most meaningful and obtainable. The individual must exert judgment in deciding how much information would be useful and must consider the difficulty and necessary cost to obtain it. Detailed information can be obtained only from detailed records and accounts, and these usually require considerable time and energy. Most farmers are content to supplement the general set of records and accounts with a few additional records such as for feed, labor, and machinery service. Many expenses of running a farm business are general and not easily assigned to individual enterprises. For individual enterprise records to be meaningful general expenses must be allocated to each of several enterprises as accurately as possible. If the allocation is not reasonably accurate, the result from the enterprise analysis may also be distorted. If information is lacking on how to allocate expenses, it may be better to omit some elements rather than guessing and obtaining distorted results.

139

In order to be meaningful, enterprise records should be kept on the accrual basis. This way any inventory change is applied during the accounting period. A decision also must be made as to what accounting period should be used. In some instances the operator may like to have an analysis on a particular lot of cattle, hogs, or broilers. In this case, the accounting period must be adjusted to fit the production span of this particular group. Most operators find it easier and more meaningful to make an enterprise analysis on a calendar or fiscal year basis. In this case accounting should be made of the inventory on hand at the beginning and the end of the period. Generally, the accounting period for the enterprise would coincide with the accounting period for the total business records. The type of bookkeeping system used by the farmer may determine the feasibility of enterprise records, the type of records to keep, and the accounting period to be used. With a detailed, computerized accounting system it may be feasible to break the enterprises down into individual lots or groups for analysis. For most enterprises keeping records on a calendar or fiscal year basis would be easier and would provide essentially the same data.

The important details to record relate to the variable cost elements. These are useful data for management purposes. In planning for short-time production periods of one year or less, the variable costs are the key considerations relative to production volume and methods. Expenses such as taxes (particularly on land and improvements), interest, rent, operator labor, and depreciation may not be useful information in analyzing a particular activity in the short run. However, for new activities some of these fixed costs may be helpful. It should be realized that over time some of the fixed costs may become variable costs. For instance, consideration of the construction of a new livestock production facility would be a variable cost at that point in time. Once the structure is built, at least part of the associated cost will be fixed cost, continuing whether the unit is used partially, completely, or not at all.

Both cash cost and opportunity cost should be considered in enterprise analysis. Even though operator labor may not have a cash cost, the income given up to use this labor in alternative enterprises may be very real. Using the operator's labor in one enterprise may be profitable, but the question should be raised and answered as to whether this labor would provide a higher return if used elsewhere in the business.

When it is desirable to allocate fixed and overhead costs for specific enterprises and activities, the following procedures have been used:

1. Allocate fixed elements on the basis of the contribution each enterprise or activity makes to gross income.
2. Allocate fixed elements on the basis of some use record or occupancy percentage.
3. Allocate on the basis of custom rates. A certain percentage of custom rates are for variable and/or fixed elements.

Some accounts can be organized to associate the fixed costs with particular enterprises or activities. The credit account, depreciation schedule, and tax bill are examples where account organization can be helpful. The allocation of overhead cost to supplementary enterprises or activities can lead to erroneous interpretations. Supplementary enterprises and activities should never be charged for feeds, labor, or other resources that have no value (zero opportunity costs) except when utilized in the supplementary enterprise or activity.

Service activity records may be very helpful in evaluating enterprises. The most useful are for labor and machinery. These were discussed briefly in Chapter 6. These records will provide information for allocating both variable and fixed costs. Service accounts can be organized to develop these costs as well as to allocate them. Many of these items are more a matter of account organization than of collecting additional information.

Enterprise records can be helpful to the manager or decision maker in several ways. First, these records can pinpoint the level of profitability and the factors affecting this profitability during the enterprise accounting period. Second, if records are kept over time, the level of performance from year to year can also aid the manager in making wise decisions relative to the future. Comparison over time may indicate how successful management performance is in maintaining or improving profitability of the individual enterprise. The operator who keeps good records will probably be proud of maintaining and comparing these records over time. This is especially true if the operator is making good financial progress or is improving the volume or performance of the enterprise.

LIVESTOCK ENTERPRISE ANALYSIS

When livestock enterprises are analyzed, some factors can be very important in assisting the decision maker in planning for the future. These factors can also point out why the enterprise was successful or why it was not as profitable as the operator planned originally. Some very important factors in most livestock enterprises are:

1. *Livestock returns per $100 feed fed* will vary with different livestock enterprises. A comparison needs to be made with other specialized farms as to the level of profitability achieved compared to competing farms. For instance, a return of $150 per $100 feed fed would be very profitable in a cattle-feeding enterprise, but would be unprofitable in a dairy enterprise since feed costs are a much higher percentage of the cost of production in cattle feeding than in dairying.
2. *Feed costs per 100 pounds of pork, beef, or milk produced* provide an indication of the cost of feed relative to the return received for the product. Since feed is one of the highly variable costs in livestock production, it provides some guidelines for determining the profitability of the enterprise.
3. *Pounds of feed per pound of product produced* is a guideline to the efficiency of the individual enterprise. These conversion figures must be compared with efficient operations of the same type to provide guidelines.
4. *Profit from the total enterprise or per unit* is a key consideration. Recognition must be given to the fact that a high profit on relatively few units may not add a large profit to the total business. In enterprise analysis there may be a compromise between the profitability per unit and that of the total enterprise. The operator may seek more volume of production even though the profit per unit may be somewhat less. He may find that the added volume contributes far more to profit than the lower profit per unit subtracts.
5. *Other factors* that might be important include such items as pigs weaned per litter, percent of calf crop, percent of death loss, eggs produced per hen, milk production per cow, pounds of gain per day per steer, gross income per dollar of expense, etc.

An example of a livestock enterprise analysis for raising swine on the Dunn farm is shown in Form 8.1. Note that on the first page income and expenses are listed in much the same form and detail as shown on the income statement for the total farm (Table 5.5). Expenses are divided into operating and fixed; operating expenses are relatively easier to tabulate than fixed expenses. Also, operating expenses are more important; failure to compute one or more fixed expenses does not make the analysis invalid. Use of variable and fixed costs have been discussed elsewhere. Most of the data appearing in Form 8.1 can be found in the tables and forms in Chapters 2-6.

Production figures and efficiency measures are shown on the second page of Form 8.1. Tabulation procedures are given in the reference column. Sources of the quantity items are production records and income and expense accounts. The measures are much the same as for the total farm analysis except that these relate to only one enterprise. It is not necessary to tabulate all measures called for to find the form useful. The most productive analyses have been discussed earlier in this chapter.

When keeping enterprise accounts, the operator should always strive toward simplifying bookkeeping detail. He may be able to use some shortcuts that will make the task easier but that will not materially influence the accuracy of the analysis. For instance, if there are two enterprises on the farm, such as hogs and beef feeding, it may be feasible to inventory the corn in individual bins or cribs and then feed one enterprise specifically from one bin. Thus it would not be necessary to keep a day-to-day record on the amount of feed going to each enterprise unless the operator felt this was desirable. If the bin contained 2,000 bushels of corn, he could allocate this amount of corn to the enterprise when the bin was emptied. If he used only part of the bin, he could take an inventory at the beginning of the period and at the end and determine the amount fed. It may be possible to allocate feed from a bin on the basis of wagon or truck loads. If the feed is purchased, it can be allocated at the time of purchase, delivery, or payment. If supplement is ordered for both hogs and cattle, this can be broken down at the time it is entered into the record book. Adjustments would then need to be made only in the inventory on hand at the end of the accounting period. At the end of the accounting period, it may be desirable to balance out to make sure that all feed is accounted for and that proper allocation was made.

CROP ENTERPRISE ANALYSIS

Crop production analysis would normally be on the basis of one production cycle or on a calendar year basis. Where there is some double-cropping, deviations from this may be feasible. In crop enterprise analysis it must be realized that land is a fixed cost to the operator. This may be on the basis of interest charge on the land if he owns it, cash rent if he rents it, or a share of the crop if he is on a crop-share lease. The manager decides what variable costs to include with the land cost to maximize profits for each production period. If the land resource is subject to erosion or other factors could create problems, it may be necessary to exercise rotation of crops over time to maximize production. Thus the decisions on crop production may be on an annual basis in some instances and on a length of rotation basis in others.

Having information on crop production costs and returns provides guidelines for deciding whether to participate in government programs. Data on potential returns over variable costs for different crops can be compared to the returns from government programs.

If two or more major crops are grown on the farm, it may be difficult to allocate certain expenses to specific crops. For instance, machinery depreciation and other ownership costs are spread over all cropland. The overhead charge for the business and often for labor as well cannot easily be allocated to specific crops or fields. The fixed land charges may also be difficult to allocate to a specific crop. If the land is valued at $400 an acre and we assume a 7% interest charge and $5.00 taxes, we would have $33 of fixed cost per year. If it is necessary to maintain 40% of this land in soil conserving crops in order to maintain productivity and row crop yields at a high level, it may not be realistic to assign this cost to all crops. A higher charge should be made to soil depleting crops than to soil conserving crops.

A comparison of production costs and returns for crops over time may be more valuable than for livestock. Yield levels can indicate both the increase in productivity and the level of management that is being provided for crop production on the individual farm.

Some of the key analysis factors in crop enterprise analysis are as follows:

1. *Yields per acre* will differ from year to year because of weather variations. Therefore, the individual operator may wish to compare his yields with the county or state average to ascertain whether they are continuing to be average, above average, or below average. Good weather may produce a higher yield in a given year; but if the county average or the state average goes up even more, this does not indicate that the level of management may be the key factor in increasing yields.

2. The *gross value of crops per acre* combines yield and price and reflects selection of crops that are most profitable. The gross value of crops per acre may be compared with land values for harvested crops such as corn, soybeans, oats, and hay. The gross value of crops should be at least 20% of the value of the land. For unharvested crops such as pasture the value would be somewhat lower. If the land value is influenced by factors other than agricultural production, the value of crops may not be at the 20% or higher level. This is a rule of thumb that may be helpful. The individual operator should check the average in his area over time to establish a guideline as to whether the gross value of crops on his farm is consistent with the value of land.

3. *Machine and power costs* are related somewhat to size of unit but also vary considerably among farms. High power and machine costs are not always associated with high yields or good profits. The operator must be sure he has adequate machinery to handle the cropping operations in a timely and efficient manner. However, excess machinery may only add to the cost and may not increase returns. What is adequate machinery on a given acreage and situation is a judgment factor. Tabulations are often made on a per rotated acre basis.

4. *Fertilizer cost per acre* should be associated with yields. If fertilizer cost per acre is higher than average for the state or area, yield should be also.

5. *Production costs per unit* are not kept in enough detail by many farmers to reveal production costs per bushel or ton. Production costs over time will give the farmer who does keep a complete set of records some guidelines as to the profitability of the individual crop. Because machinery is used in all crops, it is rather difficult to assign some of the fixed costs to individual crops. It is easier to break out variable costs such as feed, fertilizer, chemicals, harvesting costs, drying costs, etc.

An example of a crop enterprise analysis is shown in Form 8.2 for corn on the Dunn

Form 8.1. Swine Raising Enterprise Analysis

	(Includes only those items for the enterprise being analyzed.) Item	Reference (computation)	19 72	19__	19__	19__	19__
1	INCOME SUMMARY		xxx	xxx	xxx	xxx	xxx
2	Increase		xxx	xxx	xxx	xxx	xxx
3	Sales: Progeny - *raised*		24,577				
4	Breeding stock		2,656				
5	*Purchased hogs*		3,576				
6							
7	Total Sales	(3 to 6)	$30,809	$	$	$	$
8	Inventory adjustment:		xxx	xxx	xxx	xxx	xxx
9	Increase						
10	Decrease		2,968				
11	Home-consumed products		160				
12	Gross Increase	(7+9-10+11)	$28,001	$	$	$	$
13	Purchases: Feeders						
14	Breeding stock		445				
15							
16	Total Purchases	(13 to 15)	$ 445	$	$	$	$
17	Cash Increase	(7-16)	$30,364	$	$	$	$
18	Net Increase (gross income)	(12-16)	$27,556	$	$	$	$
19			xxx	xxx	xxx	xxx	xxx
20	Expenses		xxx	xxx	xxx	xxx	xxx
21	Feed: Supplements		6,610				
22	Grains		11,072				
23	Pasture		400				
24							
25	Total Feed	(21 to 24)	$18,082	$	$	$	$
26	Veterinary		564				
27	Supplies		135				
28	Marketing		201				
29	Labor, hired		2,178				
30							
31							
32	Total Operating Expenses	(25 to 31)	$21,160	$	$	$	$
33	Repairs: Equipment		230				
34	Buildings		314				
35	Taxes		57				
36	Insurance						
37							
38	Total Cash Fixed Expenses	(33 to 37)	$ 601	$	$	$	$
39	Total All Cash Expenses	(32+38)	$21,761	$	$	$	$
40			xxx	xxx	xxx	xxx	xxx
41	Depreciation: Equipment		180				
42	Buildings		752				
43	Total Depreciation	(41 to 42)	$ 932	$	$	$	$
44	Total All Expenses	(39+43)	$22,693	$	$	$	$
45			xxx	xxx	xxx	xxx	xxx
46	Net Income	(18-44)	$ 4,863	$	$	$	$
47			xxx	xxx	xxx	xxx	xxx
48	Net Cash Income	(17-39)	$ 8,603	$	$	$	$
49			xxx	xxx	xxx	xxx	xxx
50	Gross Profit	(18-25)	$ 9,474	$	$	$	$

a

Form 8.1. (continued)

	Item	Reference (computation)	19 72	19__	19__	19__	19__
51	PRODUCTION SUMMARY						
52	No. sows (incl. breeding gilts)		54				
53	No. sows bred during year		114				
54	No. farrowings during year		104				
55	No. pigs weaned		695				
56	No. hogs marketed		808				
57	Pounds pork produced		171,706				
58	No. pigs died after weaning		7				
59	No. sows died		3				
60	PHYSICAL INPUT SUMMARY						
61	Feeds: Grains and supplements (lb)		710,220				
62	Other (lb)						
63	Total feed (lb)	(61+62)	710,220				
64	Labor use: Operator and family(hr)		890				
65	Hired (hr)		1,089				
66	Total labor (hr)	(64+65)	1,979				
67	Capital investment: Livestock		$ 10,663	$	$	$	$
68	Swine equip.		$ 400	$	$	$	$
69	Swine bldgs.		$ 4,114	$	$	$	$
70	Total Capital	(67 to 69)	$ 15,177	$	$	$	$
71							
72	PRODUCTION EFFICIENCY						
73	Breeding to farrowing ratio	(54÷53)	1.1				
74	Pigs weaned/sow bred	(55÷53)	6.1				
75	Pigs weaned/sow farrowed	(55÷54)	6.7				
76	Lb. pork/sow farrowed	(57÷54)	1,651				
77	Lb. feed fed/lb pork product	(63÷67)	4.14				
78							
79	INCOME EFFICIENCY						
80	Gross income/sow farrowed	(50÷54)	$ 265	$	$	$	$
81	Gross income/hog marketed	(50÷56)	$ 34	$	$	$	$
82	Returns over feed cost	(50)	$ 9,474	$	$	$	$
83	Returns/$100 feed fed	(18÷25)	$ 152	$	$	$	$
84	Feed costs/100 lb produced	(25÷57)	$ 10.53	$	$	$	$
85	Total cost/100 lb produced	(44÷57)	$ 13.22	$	$	$	$
86	Net income/$1 expense	(46÷44)	$ .21	$	$	$	$
87	Gross income/$1 expense	(18÷44)	$ 1.21	$	$	$	$
88	Net income/sow farrowed	(46÷54)	$ 47	$	$	$	$
89	Return to labor and management	46-(70 x rate)	$ 3,952	$	$	$	$
90	Return to capital	46-(64 x rate)	$ 3,083	$	$	$	$
91	Return to capital/$1 invested	(90÷70)100	20 %	%	%	%	%
92	DEATH LOSS						
93	Percent weaning loss	($\frac{\text{Born-weaned}}{\text{born}}$)100	%	%	%	%	%
94	Percent weaning to maturity loss	(58÷55)100	.9 %	%	%	%	%
95	Percent sow loss	(59÷52)100	5.6 %	%	%	%	%
96							
97	SPECIFIC MANAGEMENT PRACTICES						
98	19___						
99	19___						
100	19___						

b

Form 8.2. _____Corn_____ Crop Enterprise Analysis

	Item	Reference (computation)	19 72	19__	19__	19__	19__
1	INCOME SUMMARY						
2	Receipts						
3	Crop sales		0				
4	Government payments						
5	Cash value, marketable crops fed		12,910				
6	Other:. _Value of landlord share_		1,952				
7							
8	Total Cash Receipts	(2 to 7)	$14,862	$	$	$	$
9	Inventory adjustment: Increase		3116				
10	Decrease						
11	Value, aftermath and by-products		346				
12	Other:						
13	Total Noncash Receipts	(9-10+11+12)	$3,462	$	$	$	$
14	Total All Receipts	(8+13)	$18,324	$	$	$	$
15							
16	Expenditures						
17	Seeds and plants		525				
18	Fertilizer and lime		2806				
19	Herbicides		125				
20	Insecticides		346				
21	Supplies						
22	Insurance		246				
23	Custom work		1,384				
24	Machinery operating costs						
25	Tractor operating costs		1,713				
26							
27	Other: _Hired labor_		865				
28							
29	Total Operating Costs	(16 to 28)	$8,010	$	$	$	$
30	Land use (interest return if owned. cash rent value if rented)		4,020				
31	Value of manure applied						
32	_Cash rent_		1,054				
33	Depreciation: Machinery						
34	Equipment		1,883				
35	Other: _Buildings_		200				
36	Total Noncash Expenses	(30 to 35)	$7,157	$	$	$	$
37	Total All Expenses	(29+36)	$15,167	$	$	$	$
38							
39	Net Income, Crop Enterprise	(14-37)	$3,157	$	$	$	$
40							
41	Net Cash Income, Crop Enterprise	(8-29)	$6,852	$	$	$	$
42							
43	Gross Profit, Crop Enterprise	(14)	$18,324	$	$	$	$
44							
45							
46							
47							
48							
49							
50							

a

Form 8.2. (continued)

	Item	Reference (computation)	19 72	19__	19__	19__	19__
51	PRODUCTION SUMMARY						
52	Number acres planted		173				
53	Number acres harvested		173				
54	Planting rate		18,000				
55	Quantity harvested or produced		17,122				
56							
57	INPUT SUMMARY						
58	Labor use: Operator, family (hr)		398				
59	Hired		434				
60	Total Labor	(58+59)	832				
61	Capital investment: Buildings		$ 2,000	$	$	$	$
62	Machinery		$	$	$	$	$
63	Equipment		$ 6,202	$	$	$	$
64			$	$	$	$	$
65	Total Capital	(61 to 64)	$ 8,202	$	$	$	$
66	Fertilizer/acre applied: N		150				
67	P		80				
68	K		60				
69	Lime						
70							
71	PRODUCTION EFFICIENCY						
72	Yield/acre	(55÷53)	99				
73							
74	INCOME EFFICIENCY						
75	Net income/acre	(39÷53)	$ 18.25	$	$	$	$
76	Net income/unit of production	(39÷53)	$ .19	$	$	$	$
77	Fertilizer, chemical cost/acre	(18 to 20)÷53	$ 16.22	$	$	$	$
78	Machinery cost/acre	(24+25+33+34) ÷53	$ 20.79	$	$	$	$
79	Net income/$1 expense	(39÷37)	$ .22	$	$	$	$
80	Gross profits/$1 expense	(43÷37)	$ 1.22	$	$	$	$
81	Returns to labor and management	39-(65xrate)	$ 2,665	$	$	$	$
82	Returns to capital	39-(58xrate)	$ 2,361	$	$	$	$
83	Returns to capital/$1 invested	(32÷15)	$ .38	$	$	$	$
84							
85	SPECIFIC PRODUCTION PRACTICES						
86	19___						
87	19___						
88	19___						
89	19___						
90	19___						
91							
92							
93							
94							
95							
96							
97							
98							
99							
100							

b

farm. Most individual entries can be found in tables and forms in Chapters 2-6. The first page of Form 8.2 shows figures for returns and costs. Variable or operating cost items are more important than fixed cost items. A meaningful analysis can be made using only variable costs. The second page shows production items and efficiency measures. Tabulation procedures appear in the reference column. It is not necessary to tabulate all measures called for to have a meaningful analysis. One of the main measurements of success of an enterprise is its profitability. The operator also should know what other operators are doing and should compare his performance with similar farms. This will give him some guidelines as to his level of efficiency under the present cost-price situation.

9

ELECTRONIC DATA PROCESSING

NATURE AND IMPORTANCE OF ELECTRONIC DATA PROCESSING

Electronic data processing, commonly known as EDP, is relatively new in the field of farm record keeping. Refinements and improvements in computer hardware and data storage have provided accountants, economists, and computer programmers new tools for developing computerized farm accounting programs.

Electronic data processing can be defined as a system of processing accounting records on an electronic computer. The amount of computer involvement varies with each EDP program. In most programs the farmer subscriber records the transaction information on prepared forms, or in a record book, and submits these to the servicing organization on a monthly, quarterly, or annual basis. The mail is the usual means of transmission. The subscriber identifies each transaction with a predetermined code number or name. The servicing organization then processes the farm data on computers that have been programmed to perform the posting, adding, subtracting, multiplying, dividing, and other accounting functions necessary to print out the desired accounting reports.

Farm record keeping must keep pace with the many technological changes in agriculture. Farmers are realizing they need a system that will handle a vast amount of data quickly, efficiently, and accurately. They want to be able to analyze their business operation through monthly reports rather than merely at the end of the year when filling out income tax forms. Forward planning is becoming a necessity rather than an exception.

As profit margins continue to shrink, farmers are concerned with total farm costs and revenues as well as whether each enterprise in the farming operation is profitable. Farmers are pinpointing the strong and weak areas of their farming operations through enterprise analyses. Livestock enterprises are being compared to each other and to grain enterprises. In some cases, lots of livestock and fields of grain are analyzed to determine the most efficient production method to be used. To do this accurately requires recording intrafarm transfers, i.e., within the farm from one enterprise to another. These transfers between enterprises, such as crediting the corn enterprise and charging the cattle enterprise with the value of the corn fed, have formerly gone unrecorded.

149

Through EDP programs farmers are receiving quarterly and monthly cash-flow reports and are thus able to exert financial control throughout the year. Farmers are keeping good records for tax purposes as well as using them to become better managers so they will not be one of the statistical numbers that left farming. Mistakes or profits today are likely to involve 800 acres of cropland rather than 80, or 150 dairy cows rather than 15. Through EDP, farmers are computerizing some of the pencil pushing. Many are realizing for the first time the value of accurate, current, detailed farm records.

Computer Functions

A farmer participating in a full EDP program usually sends his business transactions and production records to the servicing organization on a monthly basis. The agency then keypunches this information onto cards or magnetic tape to be read into the computer. The computer can be programmed to perform essentially four functions with the input information.

1. The computer sorts the varied transactions into specific categories. Expenses are sorted from receipts, and each of these may be further sorted by like transactions — i.e., fuel purchases, labor expenses, livestock receipts, crop receipts, etc.
2. The machine subtracts, multiplies, and divides as necessary to tabulate accounting reports and make management or efficiency analyses.
3. Output reports are printed according to the exact report format programmed into the computer.
4. Storing information is another function of the computer. The sorted and calculated information can be stored on magnetic tapes or disk packages for use with information received later to calculate and print out reports covering longer periods. For example, if transactions were reported by the farmer once a month, the computer could be programmed to prepare both a monthly and year-to-date summary of receipts and expenditures.

At this point a word of caution may be in order. Subscribing to an EDP program is not a cure-all for poor records, nor is it a guarantee that you will have good records. The output information is only as good as the record information that goes into the machine, or in IBM parlance, GI = GO (garbage in = garbage out). Electronic data processing subscribers still have the responsibility of recording the information accurately and regularly. However, EDP programs can act as a stimulus for keeping farm records up to date. The computer is able to relieve the subscriber of many of the more tedious aspects of farm record keeping such as posting, adding, subtracting, calculating analysis ratios, and preparing accounting reports. The computer is a tremendous timesaver; and once the information is in the machine *accurately,* an unlimited number of reports and summaries can be produced.

With time, EDP systems have grown from a simple cash-flow analysis of income and expense to information packages consisting of enterprise analysis reports, depreciation schedules, net worth statements, tax worksheets, family expense summaries, and farm business analysis reports. Together, all these reports can supply a farmer with a relatively complete physical and financial picture of his farm business operation.

Classification of Record-keeping Procedures

Record-keeping procedures are most often classified as to how the data are gathered and recorded. Thus two systems of classifications evolve:

1. Cash-basis record keeping
   a. Single-entry accounting system

   Accrual-basis record keeping
   a. Single-entry accounting system
   b. Double-entry accounting system

2. Single-entry accounting system
   a. Cash-basis accounting
   b. Accrual-basis accounting

   Double-entry accounting system
   a. Accrual-basis accounting

The similarities and differences between cash-basis and accrual-basis record keeping and between single-entry and double-entry accounting systems were discussed in Chapter 5. Clearly both classification systems are categorizing the same record-keeping concepts; they differ only in relative emphasis. One classification system cannot be described as more correct than the other; several methods may be used to record data and the procedures and information recorded vary. If these differences are not clear at this point, a review of Chapter 5 is advised.

For the purposes of this chapter a slightly different classification from the two above will be used. In classifying EDP record-keeping programs into package types, emphasis will be placed on the amount of computer involvement, subscriber recording procedures, and the type of information provided. This does not ignore the previously mentioned systems of classifying farm record-keeping programs. The distinguishing characteristics and features of cash-basis, accrual-basis, single-entry, and double-entry accounting are as much a part of EDP record keeping as conventional programs.

TYPES OF EDP PACKAGES

Several EDP record-keeping packages are available. They differ in the amount of computer involvement required, the type of external services offered, and the amount of information provided. The amount of computer involvement varies from processing individually recorded transactions on a monthly basis to processing record book summary totals on an annual basis. Some EDP packages offer such external services as income tax preparation, record-keeping assistance by area fieldmen, and farm management consultation. The information provided varies from a simple cash-flow summary to complete management information systems providing accounting and business analysis reports and forward planning options.

Year-end Farm Summary Programs

The year-end farm record summary contains the least amount of computer involvement of any EDP package. In this program the farmer keeps his records at the farm and posts transactions to them throughout the year. Then the receipt and expense, asset and liability, and production totals are entered into the computer for year-end summary and analysis. An income statement, a net worth statement, and efficiency measures usually make up part of the summary.

Year-end farm record summary programs lend themselves equally well to cash or accrual accounting procedures and in many cases the record book is so designed. Usually summary record-keeping programs are designed to employ the single-entry system of accounting because farmers generally have had very limited training in accounting and as a result find single-entry recording somewhat easier to understand.

Coded Check Programs

Coded check EDP programs require using specially printed checks and deposit slips that allow for entering a code number to identify the purchase or receipt. Since these checks are the major source of input information, this type of program also has been labeled a cash-flow program. Generally, only transactions involving cash payments or receipts are recorded. As a result, most coded check programs cannot accommodate accrual record keeping. However, some do provide for supplemental reporting of inventories and other noncheckbook type transactions.

Mail-in Programs

The mail-in system is another type of EDP package. The subscriber records each farm transaction with appropriate identification code numbers on prepared carbon paper forms. At the end of the month, quarter, or year (monthly reporting being the commonest) the subscriber mails one or more of the carbon copies of his recorded transactions to the servicing organization for keypunching and processing on the computer. The subscriber keeps a carbon copy of his recorded entries for his own records and reference.

Since many farmers use a single-entry cash recording system, most mail-in programs have been designed to employ this method, although some mail-in programs are designed for accrual record keeping and double-entry accounting procedures as well. To record transactions in a double-entry system, at least two code numbers must be entered for each transaction. The first indicated the account to be debited and the second indicates the account to be credited.

Computerized Depreciation Schedule

Computerized depreciation schedules are often maintained and updated by tax accountants and lawyers who provide tax services to farmers. Since tabulating depreciation schedules by hand is such a time consuming task, some tax professionals have computerized this part of their tax service.

Family Living Packages

Some EDP programs provide family income and expense summaries; these may also be a part of coded check or mail-in EDP packages. As before, an identification code number must be recorded for each income and expense entry. The recording forms, as well as the amount of expense breakdown and other information provided, vary with each package. In some programs only a few broad income and expense categories are provided. The more complete systems permit itemizing in detail such expenses as food, clothing, household, utilities, medical, contributions, recreation, and automobile.

One may question how this relates to the farm, but the relationship is more direct than may first appear. Profits generated from the farming operation are allocated basically in three ways: for making principal payments on present liabilities, for undertaking new investments, and for family living. It is important to keep track of expenditures in each of these areas. Outstanding liabilities and new investments are part of the farm records in most cases. However, in many conventional record-keeping programs, family expenses are often thought of as being rather insignificant; in fact, they are sometimes totally ignored. In contrast, many EDP subscribers feel this is one of the more valuable services offered. For the first time, many operators know how much of the farm income is being transferred to the family for food, household items, clothing, recreation, etc.

Specialized Programs

A final category of EDP packages could be described as specialized record-keeping programs or, in some instances, planning packages. Dairy Herd Improvement Association records and beef feedlot performance records are now computerized and are examples of specialized EDP packages, which vary from linear programming models to determine least-cost feed rations to simulation models projecting expected yields with alternative fertilization and planting levels. Forward budgets and projected cash flows are also examples of EDP planning packages.

TYPES OF FIRMS OFFERING EDP PACKAGES

Five basic types of groups offer EDP services to farmers: farm business associations, state universities, banks, farm service organizations, and private firms.

Farm Business Associations

In many states farm business associations are formed solely for the purpose of providing assistance in record keeping, tax service, and management consultation through area fieldmen. They are nonprofit, self-governing organizations with membership dues determined by the costs incurred in providing the farm record-keeping program. Farmer members serve on committees that coordinate and direct all activities at the area and state level. By working together, farm business association members are able to acquire the services of a computer processing center and can thus participate in the opportunities available through EDP. In addition to providing each member with summaries of his farm business, many farm business associations also provide comparative analysis information, thus permitting a member to compare the figures of his own farming operation with the averages of other members who have similar resources and production activities.

Universities

Universities have been forerunners in research and development of EDP programs. In many states, research has been followed by implementation of EDP services through extension programs and through personnel who work directly with the farmer in his record-keeping activities. Universities and farm business associations are unique in offering management consultation. The amount of EDP available varies from year-end computer processed reports in some states to monthly processing of the farm record information in others.

Banks

Local banks are also a source of EDP services; they offer the programs to provide more services to customers and also as a means of improving and facilitating borrower-lender relationships. Bankers want to know the present financial condition and credit needs of their farm customers and this is possible through accurate, easy-to-read farm records. If the bank has sufficient size, staff, and capital, it may be able to develop and market its own EDP program. In most cases, banks in agricultural areas are too small to undertake this project on their own; instead, they often purchase a franchise in a developed EDP program from a larger bank or company. The local bank then has the responsibility of selling the program to its farm customers. This often involves group meetings and personal visits to explain the service offered, what the program can do, how it works, what it costs, etc. The bank also has the responsibility of servicing the program. This includes handling questions about how to record specific transactions and which code numbers to use, how to correct an error in recording once it gets into the computer, how to interpret and understand the output reports, and in general, how to make better use of the EDP program. Bank employees are trained to help subscribers in these areas. Many bank EDP programs are somewhat limiting in that only cash-flow and tax data information are provided; as a result, only a minimum of business analysis information is available.

Farm Service Organizations

Another source of EDP is from farm organizations such as the Farm Bureau and Production Credit Association (PCA), which are interested in providing services to their members and have recognized farm record keeping as an area in which to do this. They have sufficient size and membership to develop EDP programs and process record information on their own computers. The Iowa Farm Bureau services its program through group meetings, often in conjunction with a local bank or vocational agriculture department, through a WATS line and through personal contact as needed. The PCA offers its program through its local office. In many cases one person at the local level is designated as a farm records coordinator to work exclusively with the record-keeping program. The PCA desire, as with banks, is to improve and facilitate the borrower-lender relationship.

Private Firms

Private firms have also encouraged better farm records and record-keeping practices by offering EDP programs. They may work exclusively in computerized farm record-keeping services for profit. In some cases, EDP services may be one of several areas within, or "products" of, a large private firm. The service may be used as part of a sales promotion program or as an enticement to attract customers to the other products the firm sells. Again, if the firm is relatively small, it may want to purchase an operational program rather than assuming the large development costs involved in offering EDP services. In other cases, where larger established firms have decided to offer EDP services, the parent company may finance the developmental and implementation costs incurred by a subsidiary firm in making the program operational.

OPERATIONAL PROCEDURES

Operational procedures vary with each type of EDP package. With year-end farm summary programs the farm record information must be recorded throughout the year in a record book kept at the farm. At the end of the year, the farmer or his area fieldman totals his record book columns for expenses, receipts, production, assets, and other farm record information. These totals are then entered on input forms (such as Form 9.1 from the Iowa Farm Business Association) to be submitted to the computer for processing. Each information box on the input form is numbered to correspond with similarly numbered column totals in the record book. This ensures that the proper column total is entered in the correct space on the input form. At the processing center the information is keypunched onto cards to be read into the computer. This type of EDP program involves a considerable amount of manual work by the farmer or his fieldman.

Code Systems

In EDP packages utilizing the computer more fully, such as the mail-in and coded check programs, each farm transaction is entered into the computer. This necessitates devising some way to indicate the nature of the transaction to the computer. The commonest method of doing this is by numerical code. This aspect of EDP programs is most foreign to subscribers. Farmers are accustomed to writing expenses or receipts under the proper column heading in a record book, as in the program previously described, rather than using a code number to indicate the nature of the entry. It is not difficult to use EDP code systems but they do require some ''getting used to.'' The code system must be understood and properly used to produce accurate output reports and information. Assigning the proper code number to each transaction entry is one of the most important responsibilities of an EDP subscriber. Some servicing organizations will, for an additional fee, code the recorded entries and thus relieve a subscriber of this responsibility.

Many different types of coding systems are used. The three-digit code system from Rec-Chek shown in Form 9.2 is one example. With this particular coding system only the commonest category headings are listed. Unused code numbers are available to the user to identify items as desired for further specification. For example, number 500 labeled ''Feeds'' could be itemized to the following feed categories:

501	Beef feed	504	Poultry feed
502	Dairy feed	505	Sheep feed
503	Hog feed		

Most EDP code systems offer categories for itemizing personal and family expenses as does this system (codes 000 through 099). This is an area of farm record keeping that often is totally ignored or treated in a lump sum fashion. Valuable information regarding consumer spending habits can be revealed by these simple personal and family categories. Again one could itemize categories of particular interest such as clothing and food:

006	Clothing—father	021	Food—groceries
007	Clothing—mother	022	Food—meat
008	Clothing—child A	023	Food—restaurant meals
009	Clothing—child B	024	Food—school lunches

Other code systems use four or more digit codes to identify specific items and transactions. The following five-digit code system from the Farm Bureau illustrates a system with more than three digits. First, it is structured to identify the type of trans-

Form 9.1. Year-End Farm Summary Input Record (courtesy Iowa Farm Business Assoc., Colo, Iowa)

ASSOC.____ COUNTY **25** FARM NO. **999** YEAR **21** ECONOMIC AREA __ TENURE __ L. LORD NO. **7** OPERATOR NO. **6**

CROP

CROP		ACRES		YIELD	
Corn	6	113	10	96	237
Corn Silage	235	20	7	15	26
Oats	7	48	11	73	27
Soybeans	8	55	12	27	30
Hay	9	39	13	3	247
Other Row Crop	236		6	7122	248 13540

ACRES			4	
Diverted	15	275		
Total Crop	16	275		
Total Rotated	17	290		
Total Farm				

CROP VALUES

Total Value of Crop Produced	6	18
Diverted Acre Payment	5	19
Total Value of Non feed Crops Raised	5	20
Gross Value of All Crop Raised on Rotated Land	6	21 **23007**
Est. Value of Own Sm. Grain for seed	4	22

LIVESTOCK PRODUCTION (Total Farm)

No. of Beef Cows	3	23	20
No. of Litters	3	26	50
No. of Milk Cows	3	27	
No. of Laying Hens	6	30	
Feed Fed to Hogs	6	247	7122
Feed Fed to Beef	6	248	13540
Feed Fed to Dairy	6	249	
Feed Fed to Poultry	6	250	

LABOR (Total Farm)

		MONTHS 2
Hired Labor	31	1
Operator Labor	32	12
Family Labor	33	1

LANDLORD LABOR

	238	1

LANDLORD (Total Farm)

LANDLORD INCOMES		5
L. L. Crop share rent	205	5701.
L. L. Cash rent Income	206	
L. L. Share Feed Crops	207	
L. L. Share Non-feed Crops	208	

BALANCE SHEET

KIND OF PROPERTY (From inventory forms)

Line		OPENING			CLOSING		
	Number	Weight	Operator	Landlord	Number	Weight	Operator	Landlord	Farm
1 LIQUID ASSETS—Cattle, far and young	34 158	40 6300	46 2420.	209	58 124	64 62000	70 41700	221	xxxx
2 Hogs, feeding	35 207	41 20700	47 3517	210	59 176	65 20490	71 4453	222	xxxx
3 Poultry	36	42	48	211	60	66	72	223	xxxx
4 Sheep or other	37	43	49	212	61	67	73	224	xxxx
5 Feeds	xxxx	xxxx	50 7000.	213 4000.	xxxx	xxxx	74 7200	225 4300	xxxx
6 Soybeans, Seeds	xxxx	xxxx	51 8000.	214 1500.	xxxx	xxxx	75 5100	226 1700	xxxx
7 Fertilizer	239	xxxx	239	244	242	xxxx	xxxx	246	xxxx
8 Cattle—Breeding Stock	38 20	44 20000	53 5500.	216 5500.	62 20	68 20000	77	228 6000.	xxxx
9 Hogs—Breeding Stock	39 25	45 11250	54 1350.	217	63 20	69 9000	78 1080.	229	xxxx
15 Total Working Assets	xxxx	xxxx	55 17450	218	xxxx	80	79 9200	230	xxxx
17 FIXED ASSETS—Farm Improvements	xxxx	xxxx	56 5300	219	xxxx	xxxx	81 5300.	231	xxxx
Buildings and Land Current Valuation	xxxx	xxxx	57 10000	220 65000	xxxx	xxxx	82 10000	232 65000	xxxx
			DEPRECIATION - Working Assets		xxxx	xxxx	83 1175.	233	xxxx
			Improvements				3000.	234	xxxx
		Oper. 5 254	Oper. 5 240				Landlord	251	
		Oper. 6 255	Oper. 6 241				Landlord	252	

OTHER ASSETS	$	Landlord	256	
LIABILITIES	$	Landlord	257	

SUMMARY OF LIVESTOCK

		DAIRY			EGGS		POULTRY			HOGS			CATTLE			Sheep or other			
		Lbs. B.F.	Value		Doz.	Value	No.	Value		No.	Weight	Value	No.	Weight	Value	No.	Weight	Value	
Op. Used in Home	84	xxxx	xxxx	86	xxxx	88	xxxx	90	2	93 400	95 80.	100	102 300.	xxxx	104	
L. L. Used in Home	85	xxxx	xxxx	87	xxxx	89	xxxx	xxxx	94	xxxx	96	xxx 1000	103	xxxx	105	
Op. Number Died	xxxx	xxxx	xxxx	xxxx	91	xxxx	xxx	97 1	xxxx	xxxx		
Op. Number Born	xxxx	xxxx	xxxx	xxxx	92	Weaned	98	99	xxxx	xxxx		

YEARLY SUMMARY OF EXPENDITURES

	Total Accumulation by page or month	Machinery and Equipment	Other than Farm Expense	Fuel and oil for Farm Use	Power & Machine Hire	Owned Truck	Auto Expense	Utilities	Crop Expense	Livestock Exp.	Fertilizers and Lime	Miscellaneous Expense	Rent	Taxes	Interest	Insurance	Permanent Improvements	Principal Pay'ts on Loans & Acc't
O OPERATOR		106 1483.	107 994.	108 1566.		109 387.	110 327.	112 1412.	111 486.	113 1668.	114 48.	115 300.	116 1128.	117 3534.	118 307.	119 334.	
L L. LORD		192 1483.	193	194		195	196	198 365.	197	199 500.	200	201 500.	202	203 186.	204 141.	
F FARM																

156

Form 9.1. (continued)

YEARLY SUMMARY OF RECEIPTS

Total Accumulation by page or month	Dairy Products Sold Lbs. B-F-W/M	Value	Eggs Sold Doz.	Value	Poultry Sold No.	Value	Corn Bu.	Value	Soybeans Bu.	Value	Other Feed Crops Quan.	Value	Non Feed Crops Quan.	Value	Hire Mach.	Gas Tax Refund	Misc. Receipts	Gov't Paym't	Farm Dividends	Other Than Farm Income	Money Rec'd from Notes, Mort. Given, Acc'ts Rec.	Mach. & Equip. Bldg. & Impr. Profit or Loss
OPERATOR	120		121		122		123 *1123.*		124 *1410.*		125		126		127	128 *216.*	129	130	131			132
L. LORD	160		161		162		163 *4354.*		164 *1273.*		165		166		167	168	169	170	171			172
FARM																						

HOG RECEIPTS

Total Accumulation by page or month	No.	Weight	Sale of Hogs Raised	Sale of Hogs Purchased	Cost	Sale of Breed Stock for C.G.	Remaining Value
OPERATOR	133 *273*	134 *60060*	135 *10811.*	136		137 *270.*	
L. LORD	173		174		175		
FARM							

CATTLE RECEIPTS

No.	Weight	Sale of Cattle Raised	Sale of Cattle Purchased	Cost	Sale of Breed Stock for C.G.	Remaining Value
138 *157*	139 *159700*	140 *5406.*	141 *4779.*	142		
176 *5406.*	177	178				

OTHER LIVESTOCK RECEIPTS

Sale of Livestock Raised	Sale of Livestock Purchased	Cost	Sale of Breed Stock for C.G.	Remaining Value	Bldg. & Impr. Mach. & Equip. Remain. Value
143	144		145		146
179	180		181		182

YEARLY SUMMARY OF EXPENDITURES

Total Accumulation by page or month	MACH. & BLDGS. BOUGHT New Machinery and Equipment	New Improvements	LABOR Taxable Wages	F.I.C.A. Tax Employer	Total Amt. Paid	Commercial FEED BOUGHT Quan.	Value	Farm Raised FEED BOUGHT Quan.	Value	Poultry BOUGHT No.	Value	Hogs BOUGHT No.	Weight	Value	Cattle BOUGHT No.	Weight	Value	Other Livestock BOUGHT No.	Weight	Value
OPERATOR	147 *2243.*	148 *734.*			149 *578.*	150 *5400.*	151		152		153 *2*	154 *800*	155 *376.*	156 *157*	157 *93400*	158 *31720.*	159			
L. LORD	183	184			185	186 *300.*	187		188		189			190			191			
FARM																				

Form 9.2. Three-Digit Code System (courtesy Rek-Chek Inc., Nevada, Iowa)

CODE	CATEGORIES	CODE	CATEGORIES
000	**PERSONAL & FAMILY**	391	SHEEP—PURCHASED—BREEDING
001	FAMILY CAR	392	SHEEP—RAISED—BREEDING
005	CLOTHING	393	FEEDER LAMBS PURCHASED
010	CONTRIBUTIONS	394	FEEDER LAMBS—RAISED
015	DIVIDENDS	398	Sheep—Wool
017	DUES	410	POULTRY—BROILERS
018	EDUCATION	420	POULTRY—LAYERS
020	FOOD	421	Layers—Eggs
030	HOME	430	POULTRY—TURKEYS
040	INSURANCE—PERSONAL		**FARM EXPENSES AND INCOME**
045	INTEREST—PERSONAL	450	ANIMAL DRUG HEALTH SUPPLIES
049	LOAN PRINCIPAL	469	BREEDING FEES
050	MEDICAL	470	CROP CHEMICALS
057	DRUGS	500	FEEDS
065	SECURITIES, SAVINGS, INVEST	540	FERTILIZERS
075	RECREATION & ENTERTAINMENT	580	FARM FUELS, LUBRICANTS
080	TAXES—PERSONAL	590	GOVERNMENT PROGRAMS
085	UTILITIES—NON-FARM	600	FARM INSURANCE
095	WAGES, SALARIES—NON-FARM	610	FARM INTEREST
	CROPS	620	IRRIGATION
100	CORN	625	LIME
105	COTTON	628	LIVESTOCK REGISTRATION
109	GRASSES & LEGUME SEED CROPS	630	MISCELLANEOUS
131	OATS & RYE	660	REFUNDS
132	PASTURE	670	RENT OR LEASE
134	PEANUTS	680	RESEARCH, TESTING
136	POTATOES	690	MANAGEMENT SERVICES
139	RICE	692	Services—Banking
142	HAY	700	REPAIRS, MAINTENANCE
152	SILAGE	730	SEED AND PLANTS
158	STRAW	738	SUPPLIES
162	SOYBEANS	740	FARM TAXES
165	TOBACCO	750	FARM UTILITIES
170	WHEAT	760	VETERINARY
200	FRUITS, BERRIES, NUTS	770	FARM WAGES, SALARIES
250	VEGETABLES	800	WORK—CUSTOM
280	FOREST PRODUCTS		**CAPITAL INVESTMENTS, LOANS**
290	NURSERY PRODUCTS	901	BUILDINGS—LIVESTOCK
	LIVESTOCK	904	BUILDINGS—MACHINERY & SHOP
301	BEEF FEEDERS—PURCHASED	907	BUILDINGS—CROP STORAGE
302	BEEF FEEDERS—RAISED	911	EQUIPMENT—DAIRY
303	BEEF PURCHASED—BREEDING	915	EQUIPMENT—FERTILIZER
304	BEEF RAISED—BREEDING	920	EQUIPMENT—HARVESTING
331	DAIRY—DAIRY CATTLE—PURCHASE	930	EQUIPMENT—FEED AND GRAIN
332	DAIRY CATTLE—RAISED	940	EQUIPMENT—MANURE HANDLING
355	Dairy—Milk	950	EQUIPMENT—PLANTING
361	HOGS PURCHASED—BREEDING	960	EQUIPMENT—TILLAGE
362	HOGS RAISED—BREEDING	970	EQUIPMENT—TRACTOR, WHEEL
363	MARKET HOGS PURCHASED	973	EQUIPMENT—TRUCKS
364	MARKET HOGS RAISED	975	EQUIPMENT—WATER SYSTEM
380	HORSES, PONIES	980	LOAN PRINCIPAL

COPYRIGHT © REC-CHEK INC. NEVADA, IOWA MAY 1966

action and, second, to indicate the type of item involved. Here is how it works. The first digit is used for *tax identification*. The numbers 1-9 are used to identify specific kinds of transactions as follows:

1 = ordinary farm income (100% taxable) or ordinary farm expense (100% tax-deductible).

2 = item purchased for resale (feeder cattle, feeder pigs, etc.) and the gross profit is taxable.

3 = sale or purchase of capital assets (breeding and dairy livestock, machinery, buildings, etc.). Sales are subject to capital gains tax.

4 = money borrowed or debt principal paid. Not subject to taxes.

5 = nonfarm income or personal nonfarm expense.

6 = nonfarm tax-deductible personal expense (medical, dental, etc.).

9 = noncash. This number is used to record intrafarm transactions.

The second digit is used for *major class identification*. The numbers 1-9 are used to identify specific classes of property or activity as follows:

0 = labor	5 = crops related
1 = feed	6 = machinery and equipment
2 = livestock	7 = land, buildings, and improvements
3 = livestock related	8 = general income and expense
4 = crops	9 = personal income and expense

The third digit is used for *specific identification* of the type of major class item indicated by the second digit. For example, if the second digit is 2 (livestock), the third digit would indicate the type. For the third digit:

1 = beef	4 = dairy	7 = sheep
2 = beef	5 = horses	8 = hogs
3 = dairy	6 = poultry	9 = other livestock

The fourth and fifth numbers are used for *individual identification* to allow for recording transactions involving individual lots of livestock or individual machinery items.

An example may make this code system clearer. Suppose a farmer buys some feeder pigs. The first digit of the code number to record this transaction would be 2 since feeder pigs are an item purchased for resale. The second digit would also be 2 to designate the item as livestock, and the third would be 8 to indicate the livestock are hogs. If the farmer desires to keep a separate record of the income and expenses charged to this group of feeder pigs, he could designate a lot number to this group by using the fourth and fifth digits of the code number. Assuming the feeder pigs are designated as group number 11, the five-digit code number for this transaction is 22811.

A coding system can be built or structured in many ways. The double-entry code system used by Pioneer Data Systems is designed to emphasize assets, liabilities, and equity accounts. The first digit of the five-digit Pioneer code indicates the type of account or transaction. For example:

1 = assets	3 = net worth	5 = expenses
2 = liabilities	4 = income	

The second digit of the code serves two purposes. With asset and liability accounts the following format is used:

> 1 = current assets or current liabilities
> 2 = intermediate assets or intermediate liabilities
> 3 = fixed assets or long-term liabilities

For income and expense transactions the second digit identifies the transaction in the following way:

Income	*Expenses*
1 = livestock sales	1 = livestock expenses
2 = crop sales	2 = crop expenses
3 = labor and services income	3 = family expenses
4 = investment income	4 = operating expenses
5 = other income	5 = other overhead expenses

The third digit carries the identification process still further. For example, intermediate assets are identified as machinery, equipment, and breeding stock while livestock sales are broken down to hog sales, cattle sales, and sales of other livestock. The fourth and fifth digits continue the breakdown by identifying breeding stock as sows, boars, cows, bulls, etc. Hog sales are broken down to feeder pigs, gilts, sows, boars, etc.

Other EDP programs use a key word code system instead of a numerical code to identify transactions. The following code word examples are from Michigan State University's TELFARM program:

> CATTLERSL = feeder cattle purchased for resale
> GASTAXS = state gas tax refund
> BORROWED = money borrowed or charged
> SUPPLEMENT = supplement and soybean meal
> RPRMACH = repairs for machinery
> TAXINCF = federal income tax paid

When evaluating EDP programs, the coding system cannot be overlooked; it is used on every transaction entered into the program. It must be well designed and contain enough flexibility to meet the needs of the farming operation.

Input Forms

The input forms used to record the code numbers, dollar amounts, and other transaction information will vary depending upon the type of firm offering the EDP service. For bank programs the usual input form is the coded check. Form 9.2 illustrates a Rec-Chek that may be used for this procedure. The bank provides a checkbook-size listing of the code categories to be used. Recording the transaction is done at the time of the purchase. The only additional responsibility, other than writing the check, is to record the code number for the item purchased. Often a check consists of several items purchased at the same store. For EDP systems where the written check is the only source of expense information, it is extremely useful to have checks, such as the one in Form 9.3, that are designed to itemize the amount and corresponding code number. If each check allows only one item to be recorded, several checks must be written at the same time to identify different types of items purchased at the same business.

For bank EDP programs, check-size deposit tickets are also supplied to record receipts and to identify the source of income by code number. See Form 9.3. Processing is usually done monthly for bank-offered EDP programs. The bank sends the checks, deposit slips, and intrafarm transfer tickets to the processing center and in most cases receives a duplicate copy of the output information received by the farmer.

Not all EDP programs use coded checks and deposit slips to record transaction information. Forms 9.4 and 9.5 illustrate input forms used to record receipts and expense payments in the Farm Bureau program. The same basic information is recorded as with the check input system but in a slightly different manner. One does not have the convenience of adding code numbers to a check to identify the purchase. Transactions on Forms 9.5 and 9.6 are actually journal entries. In this case there is not a single journal but an income journal and an expense journal. Thus the type of transaction determines on which journal sheet it is entered. Instead of posting these entries to the respective ledger accounts, we can rely on the computer to perform this accounting procedure.

When these input forms are mailed to the processing center at the end of the month, a keypunch operator will punch the code numbers and respective amounts onto cards to be fed into the computer. For example, the computer has been programmed to recognize the code number 13200 as milk sales and thus would store the amount of $335.75 in the computer area designated for the milk sales ledger account. This same procedure is repeated for every journal entry input into the computer.

Intrafarm Transfers

Intrafarm transfer information is necessary for enterprise analysis purposes; however, some EDP programs stop short of providing an opportunity to record this. If a subscriber desires to measure the true contribution of each enterprise to the total farming operation, he must record such things as the transfer of calves from a

Form 9.3. Coded Check and Deposit Slip (courtesy Rec-Chek Inc., Nevada, Iowa)

Form 9.4. Mail-in Input Record of Receipts (courtesy Farm Bureau Agricultural Business Corp., Des Moines, Iowa)

ABC FORM 3

INCOME

PAGE NO. _1_

Modern Farmer
NAME

June
MONTH OF REPORT

1-01-000
FARM NO.

LINE	DAY	CODE	ENTER-PRISE	ITEM (use Code Description or Your Own)	NUMBER	WEIGHT	AMOUNT
	1	13200		May Milk Check - Late		8071	335 75
	17	12800	HOG	Butchers to Producers @ 19.25	21	4494	865 10
	20	12800	HOG	Butchers @ 18.75	11	2530	474 38
	15	13200		June 15 Milk Check		8641	343 91
	30	13200		June 30 Milk Check		6806	262 03

Form 9.5. Mail-in Input Record of Expenses (courtesy Farm Bureau Agricultural Business Corp., Des Moines, Iowa)

CONFIDENTIAL

ABC FORM 4

PAYMENTS

Modern Farmer
NAME

June
MONTH OF REPORT

FARM NO. 1-01-000

LINE	DAY	CODE	ENTER-PRISE	ITEM (use Code Description or Your Own)	NUMBER	WEIGHT	AMOUNT
	2	15200	F 1	Plow-down Fertilizer		12,500	475 14
	28	11700		Dairy Cow Feed		500	41 75
	3	11700		Cattle Protein Feed		3,000	151 85
	14	10100		Labor - Craig Burns			390 69
	12	13800		Vet. - Supplies			66 45
	8	13800	H 0 6	Vet. - Vaccinating Pigs			141 70
	15	10100		Milking - Joe Jones			75 00
	13	11700	H 0 6	Hog Feed		2,000	109 62
	5	16700		Planter Repairs			86 97
	30	15200	F 1	Starter Fertilizer		6,000	264 06
	21	11700		Chick Feed		300	16 25
	15	16700		Repair Bill at Charley's Place			114 85
	31	15200	F 1	Fertilizer for side-dress		8000	312 08
	26	10100	F 1	Dick Brown - Cultivating			80 00

Form 9.6. Supplemental Data Entry Sheet with Provision for Correction of Past Errors

(courtesy Pioneer Data Systems, Inc., Des Moines, Iowa)

OWNER CODE: PSF
OWNER NAME: Pleasant View Stock Farm
MONTH: June

⊗ PIONEER™ DATA SYSTEMS

TRANSACTION DATE MONTH	DAY	REFERENCE	AMOUNT	DESCRIPTION (OR) QUANTITY	UNIT	DEBIT TO ACCOUNT	ENTERPRISE	CREDIT FROM ACCOUNT	ENTERPRISE
6	2	014	11,408 00	50 -	Head	41315	LT4	11510	PSF
6	11	015	1,450 15	1450 15	Bushel	42025	COM	11610	PSF
6	19	016	1,496 80	8 -	Head	41315	LT5	11510	PSF
6	27	017	250 00	10 -	Head	41315	EST	11520	PSF
6	8	418	587 18	10 -	Head	11520	PSF	41315	WST
6	10	419	60 00	1 -	Head	41315	LT5	11510	PSF
6	1		9 00	15 -	Bales	51170	BRD	43050	HAY
6	1		3,000 00	3,000 -	Bushel	51130	LT5	42020	FED
6	1		120 00	120 -	Bushel	51130	BOR	42020	FED
6	30		9 00	15	Bales	42055	HAY	11640	PSF
6	30		3,120 00	3,120	Bushel	42025	FED	11610	PSF

— MAKE CORRECTING ENTRIES IN AREA BELOW.

5	18	400	1 85	Correction		53010	FAM	53040	FAM

beef raising enterprise to a beef feeding enterprise or the transfer of corn from the corn growing enterprise for feed in the beef enterprise. The beef raising and corn growing enterprises should be credited for the value of the product transferred, and the beef enterprise should be charged for this amount just as if it had been a cash purchase.

Several methods are available for handling intrafarm transfer information for EDP programs. In most cases supplemental or intrafarm journal sheets are required to record this information, but not always. The five-digit Farm Bureau code system illustrated earlier is structured to facilitate coding intrafarm or noncash transfers by merely coding the first digit of the code with the number 9. No supplemental entry sheets are required. For EDP programs where this feature is not part of the code system, entry forms such as the one from Pioneer Data Systems (Form 9.6) are required. For bank programs, check-size slips are often provided to record these entries.

Correcting Errors

At the bottom of Form 9.6 is an area for correcting errors in transactions coded in prior months. In this example, reference number 400 indicates that this entry corrects a previous month's family expense entry numbered 400. Coding errors such as this are not uncommon, especially for new subscribers, and they are difficult to detect. The computer can be programmed to do some error checking such as for illegitimate codes and noncoded entries. However, no matter how complete and comprehensive the EDP program, careful checking of the coded entries by the subscriber cannot be overemphasized. The coded entries should be double checked for correct code, dollar amount, and any other information required to indicate the nature of the transaction before the records are sent to the processing center. Output reports should be checked with bank statements and any other record information available to ensure that an incorrect number has not been submitted to the computer. Keypunching of the record information at the processing center is also subject to human error. By checking the output reports carefully, possible errors arising in this way can be detected. Once error-free information reaches the computer, there is virtually 100% certainty of receiving accurate information in return.

Depreciable Assets

Recording of new capital purchases and changes in depreciable assets requires much detailed information, as Form 9.7 from the PCA Agrifax program indicates. In this area computers eliminate much work. Each machinery item, depreciable breeding animal, and building is assigned a code number to maintain its identity in computer storage and calculations. When an animal is sold or a machine is traded, its code number is deleted and replaced by the code number assigned to the replacement animal or new machine, as the case may be. Any item on the schedule may be depreciated on a straight-line basis or on any of the accelerated methods. When a farmer has only a pencil and a scratch pad to work with in figuring his depreciation schedule, human tendencies often guide him to a straight-line depreciation basis. On the other hand, when programmed, the computer can calculate double-declining-balance depreciation just as fast as straight-line depreciation.

Inventories

Some EDP programs provide net worth statements periodically and at the end of the year; this requires inventory information such as illustrated on Form

Form 9.7. Record of New Capital Purchases and Changes in Depreciable Assets (courtesy Federal Intermediate Credit Bank, Omaha, Nebr.)

FORM 7

AGRIFAX ®
AGRIFAX SYSTEMS

DEPRECIATION SCHEDULE
FOR MACHINERY
IMPROVEMENTS
AND LIVESTOCK

NAME Robert Sample
FARM NO. 02-06-01-05432

▲ 0 / 1 7 / ▲ FISCAL YEAR BEGINNING (MO.-YR.)

▲ / / ▲ OWNER OR BUSINESS CODE

	IDENTIFICATION INFORMATION		DATE ACQUIRED			COST INFORMATION				DEPRECIATION INFORMATION				
LINE	ITEM DESCRIPTION USE NO MORE THAN 16 SPACES WRITE WITH BALL-POINT PEN	IDENTI-FICATION NUMBER	MO.	YR.	NUMBER OR UNITS	BALANCE FOR DEPRECIATION PLUS SALVAGE OF ITEM(S) TRADED IN	CASH PAID OR PAID TO BOOT	N OR U	YEARS LIFE	INVESTMENT CREDIT	OPTIONAL ADDITIONAL 1ST YEAR DEPRE-CIATION	EFFECTIVE SALVAGE VALUE	METHOD DEPR. 0-NON 1-SL 2-DDB 3-SD 4-DB	DEPRECIATION PRIOR TO JAN. 1, 19
	2	3	4	5	6	7	8	9	10	11	12	13	14	15
1	JD 4020 TRACTOR	70011	1	70	1	4000.00	6000.00	N	10	700.00			2	
2	CORN PICKER	70012	1	64	1		500.00	U	10				1	250.00
3	CATTLE OILER	70013	1	69	1		209.00	N	5	4.88			1	92.20
4	HEREFORD BULL	10015	4	68	1		500.00	U	7			50.00	1	219.20
5	HEREFORD COWS	10016	1	68	20		4200.00	U	7			1000.00	1	1200.80
6														
7														
8														
9														
10														
11														
12														
13														
14														
15														
16														

AG 207 REV. 8-70

PUNCHED _____ VERIFIED _____

166

Form 9.8. Inventory Information Form (courtesy Federal Intermediate Credit Bank, Omaha, Nebr.)

AGRIFAX® AGRIFAX SYSTEMS

FORM 6
ANNUAL INVENTORY OF ASSETS and LIABILITIES

NAME Robert Sample
FARM NO. 02-06-01-5432

01 71 FISCAL YEAR BEGINNING (MO. - YR.)

1 1 OWNER OR BUSINESS CODE

LIVESTOCK

	Description	Number	$/Hd.	$ Value
101	HOLSTEIN COWS	31		8525 00
102	BRED HEIFERS	14		2800 00
103	OPEN HEIFERS	12		1440 00
104	HOLSTEIN STEERS	11		1320 00
105	CALVES.	25		1000 00
106	LAYING HENS	4900		1372 00
107				
108				
109				
110				
111				
112				

FEED AND GRAIN ON HAND

	Description	Quantity	Unit	$ Value
201	CONCENTRATE	4	TON	240 00
202	CORN	2800	BU	2800 00
203	OATS	1600	BU	1040 00
204	HAY	80	TON	1600 00
205	STRAW	30	TON	300 00
206	HAYLAGE	150	TON	1500 00
207	CORN SILAGE	200	TON	1200 00
208				
209				
210				
211				
212				

OTHER ASSETS

		Description	Quantity	Unit	$ Value
49	301	FERTILIZER			
	302				
	303	A/R DESCRIBE RECEIVABLE CUSTOM WORK			1002 00
	304	A/R DESCRIBE RECEIVABLE			
	305	A/R DESCRIBE RECEIVABLE			
01	306	FARM CHECKBOOK			796 00
71	307	SAVINGS			
	308				
70	309	PCA & FLB STOCK			3685 00
	310	STOCK			2424 00
71	311	CASH VAL. INS. (LIFE)			500 00

		Description	Number	Unit	$ Value
28	312	MACHINERY			38000 00
	313				
	314	AUTO			2600 00
	315				
	316				
	317				

REAL ESTATE

		Description	Number	Unit	$ Value
	401	FARMLAND			102500 00
	402				
	403				
	404				

SHORT TERM OBLIGATIONS (NOTES AND UNSECURED DEBTS)

		Description		$ Value
05	501	PCA		64742 12
	502	CCC		1269 00
	503			
	504			
	505			

ACCOUNTS PAYABLE (FEED, GAS, BILLS, ETC.)

		Description		$ Value
	506	FARMERS CO-OP		1200 00
	507	VETERINARY		300 00
	508			
	509			
	510			
	511			
	512			

REAL ESTATE MORTGAGES AND CONTRACTS

		Description		$ Value
01	601	FED LAND BANK		30000 00
04	602	BANK LOCAL BANK		5000 00
03	603	INS. CO		
08	604	INDIVIDUAL		
	605			
	606			
	607			
	608			

801	TOTAL ASSETS	176644 00
802	TOTAL DEBTS	102511 12
803	NET WORTH	74132 88

AG 206

PUNCHED ______ VERIFIED ______

9.8 from Agrifax. The amount of cash on hand must be determined by reconciling the latest bank statement; measuring and valuing the crop and livestock inventory; determining accounts receivable and payable; and valuing land, buildings, and equipment.

OUTPUT REPORTS

Having discussed the input forms and procedures used in EDP programs, let us take a look at some of the financial output reports available.

Cash-Flow Reports

Cash receipt and expense reports are usually printed monthly or quarterly in most EDP programs. As with Form 9.9 from the Farm Bureau, the amounts for the current month and the year to date are usually printed for comparative purposes as well as to eliminate the need for summing the totals from previous months to determine the current-year figure. Form 9.10 from Pioneer illustrates an alternative type of cash-flow report. In this case, the report presents information relating to one month only. Each cash journal entry for the month is printed including the date, reference, payee, enterprise, description, and amount to aid in identifying the cash flow. In this EDP program, the intrafarm transfers for the month are also printed, even though they have no effect on the cash-flow analysis.

Cash-Flow Performance Report

Not only is it important to trace the movement of cash into and out of the farm business throughout the year but also to develop a plan for this cash movement. Developing projected cash-flow budgets is a first step in exercising closer financial control of the farm business operation. Form 9.11 from Agrifax illustrates the type of EDP report that can be produced to give a picture of the actual cash flow of the farm operation in relation to the planned cash flow. The actual amount, the budgeted amount, and the difference between the two are shown for both the current period and the year to date. This report can be very useful in analyzing the cash-flow position of the farm business operation. With this concise cash-flow picture, areas that are exceeding or falling short of established goals can be readily located.

Many EDP programs provide a tax worksheet during October or November listing all receipts and expenses for the year as an aid to subscribers in planning purchases and sales during the remaining months of the year. This is especially helpful to farmers paying income tax on a cash basis. Many cash-basis subscribers indicate this is one of the most valuable reports they receive from their EDP program.

Income Statement and Depreciation Schedule

Forms 9.12 and 9.13 from the Farm Bureau illustrate year-end reports designed to help a farmer in tax reporting. In some cases, the reports are designed to be attached to the income tax forms in support of the key total figures that are entered. Nearly all depreciation schedules are of the type shown. The basic format usually includes totals for livestock, machinery and equipment, and buildings and improvements.

Net Worth Statement

Form 9.14 from Agrifax illustrates the format of an EDP net worth statement. Not all EDP packages provide this report or information, although some programs provide it as often as monthly. In addition to the ending net worth information, the Agrifax report also includes a brief listing of assets, liabilities, and net worth at the beginning of the year. Thus some of the basic trends that have occurred during the past year can be identified. The statement at the bottom of the report is to be signed by the subscriber after checking that the information correctly reflects his financial position. This indicates to the subscriber's local PCA that the information is accurate and permits use of this report in evaluating his financial position and credit needs for the coming year.

Farm Business Analysis Reports

In addition to financial statements, the more complete EDP packages also provide farm business analysis reports. Total farm analysis reports highlight income, expenses, and farm production, calling a farmer's attention to key analysis figures that give some indication of the efficiency and quality of management in the farming operation. Form 9.15 from the Iowa Farm Business Association illustrates a total farm business analysis report; however, some EDP programs do not supply management information of this type. Other EDP programs supply guideline ratios for livestock returns per $100 feed fed, power and equipment investment per crop acre, current assets to current liabilities, and total liabilities to total net worth. Some EDP programs offered by universities and farm business associations often provide comparative analysis figures as contained in Form 9.15. As a result, each subscriber can compare each aspect of his own farming operation with average farm figures of other subscribers. Comparative analysis information can be useful but it also can be misleading. It can indicate to a subscriber how his farming operation compares to others; however, so many varied types of farming operations are represented in averages that the figures can become nearly meaningless. It may be more useful to compare all hog operations or all cattle operations rather than all farming operations. At the same time, it should be realized that comparative analysis figures do not tell why variations occur from year to year.

Enterprise Analysis Reports

Electronic data processing programs also have the capability of producing enterprise analysis reports accurately and easily if the required information has been recorded. This would include intrafarm transfers and a detailed itemization of expenses and receipts to the respective enterprises. Form 9.16 from Pioneer analyzes an enterprise where corn was fed to livestock, while Form 9.17 from Agrifax is an analysis of a dairy enterprise. EDP programs permit all kinds of enterprise analysis possibilities. Major crop and livestock enterprises can be analyzed, as well as lots of livestock or fields of crops. For example, a farmer may feed hogs in confinement, in drylot, or on pasture on his farm and want to know if one method is more profitable than another. Each of the three hog production methods could be treated as enterprises, for this analysis. Some EDP subscribers consider a large farm machine such as a combine as an enterprise to determine its contribution to the farming program. This is especially useful where the machine is used for custom work and thus generates cash income to the farm operation. A word of caution is needed here in regard to enterprise analysis reports. It may sound like a snap to obtain this information from an EDP program. In a way this is true since it takes the computer only a few extra seconds to print out the report; but the report is only as good as the information recorded on the input forms. To accurately analyze the three hog producing methods, all feed going to each group of hogs should be weighed,

Form 9.9. Cash Receipt and Expense Form (courtesy Farm Bureau Agricultural Business Corp., Des Moines, Iowa)

Farm Bureau Agricultural Business Corp.
507 10th Street
Des Moines, Iowa 50309
Phone (515) 282-8171

CASH RECEIPTS AND PAYMENTS

Freddie Computer
Farm Records Service

Name: Modern Farmer
Soc. Sec. No. 123-45-6789
Date 12-31-70
Farm No. 1-01-000
Page 1

CASH FARM RECEIPTS			CASH FARM PAYMENTS		
Kind	Current Period	Year to Date	Items	Current Period	Year to Date
1 Cattle	665 78	2 919 96	Labor hired	235 50	1766 65
2 Beef calves			Repairs, maintenance	316 56	688 46
3 Sheep			Interest	25 40	2300 28
4 Swine	832 11	4 691 16	Rent of farm, pasture		
5 Poultry			Feed purchased	583 84	1763 73
6 Dairy products	239 15	2 706 05	Seed, plants purchased		323 06
7 Eggs			Fertilizers, lime		1746 08
8 Wool			Machine hire	196 71	757 65
9 Cotton			Supplies purchased	8 34	205 63
10 Tobacco			Breeding fees		
11 Vegetables			Veterinary, medicine	30 68	102 86
12 Grain	893 87	3 789 99	Gasoline, fuel, oil	116 11	517 57
13 Fruits and nuts			Storage, warehouseing		
14 Other (specify):			Taxes		1055 92
15 Bees			Insurance		204 96
16 Hay			Utilities	49 46	548 67
17 Honey			Freight, trucking	23 45	303 76
18 Wood & Lumber			Conservation expenses		
19 Seeds Sold			Retirement plans, etc.		
20 Other Livestock Sold			Other (specify):		
21 Other Products			Auto (Farm Share)	16 81	304 00
22 Machine work			Truck	19 48	687 53
23 Patronage dividends, Refunds & Rebates		89 65	Advertising		
24 Per-unit retains			Poultry & Other Lvstk. Purchased		
25 Agricultural program payments:			Commissions, Yard Fees, etc.		
26 (1) Cash	658 00	658 00	Misc. Livestock Expense		19 84
27 (2) Materials and services			Office Supplies	15 25	76 80
28 Commodity Credit loans under election		3 677 89	Mgmt., Records, Tax, Legal		75 00
29 Federal gasoline tax credit		42 67	Bank Charges, Org. Dues		13 28
30 State gasoline tax refund		74 62	Meetings, Travel (Reimb.)		65 00
31 Other (specify):			Other Misc.		
32 Total Ordinary Farm Income	3 288 91	18 649 99	Total Ordinary Farm Expenses	1 637 59	13526 63
33 Total Sales—Purchased for Resale	7 063 02	16 690 79	Total Payments—Purchased for Resale	850 00	3534 00
34 Breeding & Dairy Lvstk. Sold		853 29	Breeding & Dairy Lvstk. Purchased		753 50
35 Machinery & Equip. Sold			Machinery & Equip. Purchased		1243 00
36 Bldg., Tile, Fence etc. Sold			Bldg., Tile, Fence etc. Purch.		
37 Total Money Borrowed		16 810 15	Total Debt Principal Paid	824 00	5300 00
38 Total Personal Income		94 75	Total Personal Expenses		1337 26
39			Total Personal Tax Deductible Expenses	38 38	812 61
40 TOTAL CASH RECEIPTS	10 351 93	53 098 97	TOTAL CASH PAYMENTS	3 349 97	26507 00

FOR PERSONAL USE ONLY

Farm Checkbook Balance
Beginning of month $ ___ 1,869.13
Net Cash Income (Expense) ___ 7,001.96
Adjustments ___
Farm Checkbook Balance
End of Month ___ 8,571.09

FREDDIE COMPUTER

FARM RECORDS SERVICE

Use this Form to Estimate Income
For Tax Purposes (Cash Method)
Ordinary Farm Income (Year-to-date) ___ 18,649.99
Ordinary Farm Exp. (Year-to-date) ___ 13,526.63
Net ___ 5,123.36

Plus Gross Profit on sale of Items
Purchased for Resale ___ 6,796.19

Plus Estimated taxable income from
Sale of Breeding & Dairy & other
Cap. Assets (Schedule D) ___ 426.00

Total estimated gross income ___ 12,345.55
Minus estimated depr. expense ___ 2,248.00

Estimated taxable income year-to-date ___ 10,097.55

170

Form 9.10. Cash-Flow Analysis Report (courtesy Pioneer Data Systems Inc., Des Moines, Iowa)

PLEASANT VIEW STOCK FARM
CASH FLOW ANALYSIS
FOR THE MONTH ENDED 6-30-70

RECEIPTS

DATE	REFERENCE	PAYOR	ENTERPRISE	ACCOUNT	DESCRIPTION -OR- QUANTITY UNIT	UNIT PRICE	AMOUNT
6-30-70	718	TOM LOWRY	BREEDING CATTLE	STOCK RENT RCVD	FULL RENT		$150.00
6-11-70	015	CO-OP ELEVATOR	COMMERCIAL CORN	CORN SALES	1,450.15 BUSHEL	1.23	$1,783.69
6-22-70	017	CLINE BECKERY	EAST SIDE	LIVESTOCK SALES	10.00 HEAD	56.40	$564.00
6-04-70	014	MIDWEST PACKING CO	LOT4	LIVESTOCK SALES	50.00 HEAD	337.98	$16,898.76
6-16-70	016	MIDWEST PACKING CO	LOT5	LIVESTOCK SALES	8.00 HEAD	351.93	$2,815.40
6-16-70	016	MERLE BROWN	MACHINERY	CUSTOM WORK INC	CUSTOM WORK		$250.00

TOTAL CASH INFLOW - $22,461.85

EXPENDITURES

DATE	REFERENCE	PAYEE	ENTERPRISE	ACCOUNT	DESCRIPTION -OR- QUANTITY UNIT	UNIT PRICE	AMOUNT
6-17-70	429	FARM CENTER	BREEDING CATTLE	COMMISSIONS	TRANSFER FEE		$2.50
6-23-70	434	INTERSTATE TRUCKING	BREEDING CATTLE	TRUCKING EX.			$8.00
6-15-70	424	PIONEER	COMMERCIAL CORN	SEED PURCHASES	6.00 BUSHEL	20.00	$120.00
6-11-70	420	LEES	FAMILY	CLOTHING	WORK CLOTHES		$12.12
6-15-70	425	CHURCH	FAMILY	CONTRIBUTIONS			$20.00
6-17-70	430	DR NELSON	FAMILY	DENTAL EXPENSE			$35.00
6-02-70	414	SAFEWAY	FAMILY	GROCERIES	GROCERIES		$28.41
6-30-70	436	DAHLS	FAMILY	GROCERIES			$16.17
6-03-70	415	SEARS	FAMILY	HOUSEHOLD EX.	SWITCH		$4.07
6-15-70	426	IOWA MUTUAL	FAMILY	LIFE INSURANCE	LIFE INSURANCE		$37.00
6-15-70	426	DAHLS	FAMILY	MISC.-FAMILY EX	CASH		$10.00
6-16-70	427	R&LT	FAMILY	MISC.-FAMILY EX			$3.00
6-16-70	428	IOWA TELEPHONE	FAMILY	PHONE			$16.14
6-19-70	431	JONES IMPLEMENT	HAY	MISC.-CROP EX.	BALER TWINE	1.50	$20.75
6-15-70	427	BILL WESLYAN	LOT1	COMM. FEED EX.	60.00 TONS		$90.00
6-10-70	419	THOMAS RENDERING	LOT5	MISC.-LVSK.-EX.	DEAD CALF		$3.00
6-11-70	421	CO-OP	MACHINERY	FUEL	150.00 GALLON	.28	$42.25
6-12-70	422	WESTERN AUTO	MACHINERY	HAND TOOLS	WRENCH		$1.03
6-05-70	416	CO-OP	MACHINERY	LUBRICANTS	GREASE		$2.60
6-22-70	433	HARRYS WELDING	MACHINERY	MISC.-OVERHEAD			$3.09
6-20-70	432		TOTAL BUSINESS	SUPPLIES	SUPPLIES		$6.48
6-26-70	435	HARRY WALTERS	TOTAL BUSINESS	LABOR	MISCELLANEOUS LABOR		$12.50*
6-30-70	417	HOME STATE BANK	TOTAL BUSINESS	MISC.-OVERHEAD	SAFE DEPOSIT BOX		$4.00
6-08-70	41A	LLOYD BAKER	WEST SIDE	HOGS PURCHASED	10.00 HEAD	58.72	$587.18

TOTAL CASH OUTFLOW - $1,081.49

INTRA-BUSINESS TRANSFERS

DATE	ENTERPRISE -DEBIT-	ACCOUNT	ENTERPRISE -CREDIT-	ACCOUNT	DESCRIPTION -OR- QUANTITY UNIT	UNIT PRICE	AMOUNT
5-16-70	FAMILY	53010	FAMILY	53040	CORRECTION	1.00	$1.85
6-01-70	LOTS	51130	FEED CORN	42020	3,000.00 BUSHEL	1.00	$3,000.00
6-01-70	BOARS	51130	FEED CORN	42020	120.00 BUSHEL	1.00	$120.00
6-03-70	BREEDING CATTLE	51170	HAY	42050	15.00 BALES	.60	$9.00
6-03-70	LOT4	41315	TOTAL BUSINESS	11510	50.00 HEAD	228.16	$11,405.00
6-10-70	LOT5	41315	TOTAL BUSINESS	11510	1.00 HEAD	60.00	$60.00
6-19-70	LOT5	41315	TOTAL BUSINESS	11510	8.00 HEAD	187.10	$1,496.80
6-24-70	EAST SIDE	41315	TOTAL BUSINESS	11520	10.00 HEAD	25.00	$250.00
6-30-70	FEED CORN	42025	TOTAL BUSINESS	11610	3,120.00 BUSHEL	1.00	$3,120.00
6-11-70	COMMERCIAL CORN	42055	TOTAL BUSINESS	11610	1,450.15 BUSHEL	1.00	$1,450.15
6-30-70	HAY	11520	TOTAL BUSINESS	11640	15.00 BALES	.60	$9.00
6-08-70	TOTAL BUSINESS	11520	WEST SIDE	41315	10.00 HEAD	58.72	$587.18

NET CASH INFLOW - $21,380.36

* UNITS RECORD NOT UPDATED BECAUSE QUANTITY AND UNIT NOT PROPERLY CODED

171

CASH FLOW PERFORMANCE REPORT

	CURRENT MONTH MARCH			YEAR-TO-DATE JANUARY THRU MARCH		
	PROJECTED	ACTUAL	DIFFERENCE	PROJECTED	ACTUAL	DIFFERENCE
INCOME						
CURR FARM REC						
BEEF CATTLE				4,250	4,740	490
CROP	9,000		9,000-	58,000		58,000-
CORN		4,015	4,015		6,534	6,534
HAY					178	178
FLAX		1,081	1,081		7,795	7,795
REFUND					46	46
GAS TAX					553	553
AGR PROG PAY						
CUSTOM					50	50
MACHINE					338	338
OTHER						
MISCEL		80	80		80	80
BEEF CATTLE					203	203
TOTAL	9,000	5,176	3,824-	62,250	20,517	41,733-
MONEY BORROWED					15,596	15,596
TOT CASH AVAIL	9,000	5,176	3,824-	62,250	36,113	26,137-
EXPENSES						
WAGES	300	209	91-	900	498	402-
SOC SECURITY					227	227
REPAIR	500	719	219	1,000	2,012	1,012
INTEREST		2,648	2,648	6,100	9,379	3,279
FEED	100	1,139	1,039	300	1,456	1,156
SEED-PLANTS		1,234	1,234		2,991	2,991
FERTLZER-LME						
CUSTOM		18	18		18	18
SUPPLIES		66	66	200	68	132-
BREEDING				50		50-
VET AND MED		49	49		49	49
PETROL					760	760
TAXES				5,000	3,915	1,085-
INSURANCE	300	471	171	600	471	129-
UTILITIES		126	126		379	379
LEASE						
MARKTNG						
AUTO EXPSE					93	93
TRUCK					250	250
OTHER		11	11	100	617	517
BEEF CATTLE					12,280	12,280
TOTAL	1,200	6,690	5,490	14,250	35,463	21,213
FAMILY LIVING	500	500		2,000	3,150	1,150
PUR DEP. CAPTL	500	16,913	16,413	500	16,913	16,413
DEBT PRIN PAY				4,800	4,768	32-
TOTAL	1,000	17,413	16,413	7,300	24,831	17,531
TOT EXPNDITURES	2,200	24,103	21,903	21,550	60,294	38,744
PCA PAYMENTS		2,417	2,417		15,045	15,045
PCA ADVANCES	2,200	7,000	4,800	21,850	36,110	14,260
PCA BALANCE	103,331	151,797	48,466	103,331	151,797	48,466

Form 9.12. Depreciation Schedule to Aid in Tax Reporting (courtesy Farm Bureau Agricultural Business Corp., Des Moines, Iowa)

Farm Bureau Farm Records
507 10th Street
Des Moines, Iowa 50309
Phone (515) 282-8171

FARM INCOME AND EXPENSES

Farm Records Service
For Farm Bureau Member

Name	Soc. Sec. No.	Date	Farm No.	Page
Modern Farmer	123-45-6789	12-31-70	1-01-000	1

Part I — Farm Income – Cash Receipts and Disbursements Method
Do not include sale of livestock held for draft, breeding, or dairy purposes, report such sales on Schedule D.

Sales of Purchased Livestock and Other Items Purchased for Resale

a. Description	b. Amount received	c. Cost or other basis	d. Profit (or loss)
Beef Cattle	12487 43	7 328 60	5 158 83
Sheep	1306 25	1 172 00	134 25
Hogs	2897 11	1 394 00	1 503 11
TOTAL	16690 79	9 894 60	6 796 19

FARM BUREAU FARM RECORDS

ATTACH TO IRS FORM 1040F

Transfer only total profit (or loss) of Sales of Items purchased for resale to line 3d 1040F. Transfer only total on line 26 & line 49 to line 26 & line 49 on 1040F.

Sales of Market Livestock and Produce Raised and Held Primarily for Sale and Other Farm Income.

Kind	Quantity	Amount
4 Cattle		$ 2 919 96
5 Beef calves		
6 Sheep		
7 Swine		4 691 16
8 Poultry		
9 Dairy products		2 706 05
10 Eggs		
Wool		
12 Cotton		
13 Tobacco		
14 Vegetables		
15 Grain		3 789 99
16 Fruits and nuts		
17 Other (specify):		
Bees		
Hay		
Honey		
Wood & Lumber		
Seeds Sold		
Other Livestock Sold		
Other Products		

OTHER FARM INCOME

18 Machine work		
19 Patronage dividends, Refunds & Rebates		89 65
20 Per-unit retains		
21 Agricultural program payments:		
(1) Cash		658 00
(2) Materials and services		
22 Commodity Credit loans under election (or forfeited)		3 677 89
23 Federal gasoline tax credit		42 67
24 State gasoline tax refund		74 62
25 Other (specify):		
26 Add lines 4 through 25		$ 18 649 99

Part II — Farm Deductions **F**
Do not include personal or living expenses not attributable to production of farm income, such as taxes, insurance, repairs, etc., on your dwelling.

Items	Amount
29 Labor hired	$ 1 766 55
30 Repairs, maintenance	688 46
31 Interest	2 300 28
32 Rent of farm, pasture	
33 Feed purchased	1 763 73
34 Seed, plants purchased	323 06
35 Fertilizers, lime	1 746 08
36 Machine hire	757 65
37 Supplies purchased	205 63
38 Breeding fees	
39 Veterinary, medicine	102 86
40 Gasoline, fuel, oil	517 57
41 Storage, warehousing	
42 Taxes	1 055 92
43 Insurance	204 96
44 Utilities	548 67
45 Freight, trucking	303 76
46 Conservation expenses	
47 Retirement plans, etc. (other than your share – See separate instructions)	
48 Other (specify):	
Auto (Farm Share)	304 00
Truck	687 53
Advertising	
Poultry & Other Lvstk. Purchased	
Commissions, Yard Fees, etc.	
Misc. Livestock Expense	19 84
Office Supplies	76 80
Mgmt., Records, Tax, Legal	75 00
Bank Charges, Org. Dues	13 28
Meetings, Travel (Reimb.)	65 00
Other Misc.	
49 Add lines 29 through 48	$ 13 526 63

173

Form 9.13. Net Worth Statement (courtesy Farm Bureau Agricultural Business Corp., Des Moines, Iowa)

1969 DEPRECIATION SCHEDULE

FARM BUREAU ACCOUNT NO. 1-01-000   MODERN FARMER   123-45-6789   PAGE 01

CODE	DESCRIPTION	N/U	DATE ACQ MO-DA-YR	BASIS M 20	MONTHS 1ST YR	I YRS C LIFE	CASH PAID	ADJ BASIS	COST BASE	SALV	START BALANCE	DEPR	END BALANCE	SOLD MO-DA	LAST MO
32101	BLK ANGUS BULL	0010	04-24-66	SL N	8.0	N 5.00 N 4	$ 225		$ 225		$ 105	$ 45	$ 60		
32104	MIXED HEIFERS	0040	06-02-63	SL N	7.0	N 8.00	426		426		129	54	75		
32801	GILT	0010	02-26-67	SL N	10.0	N 3.00	98		98		38	38**	38**	01-02	.0
32855	NASH HAMP BOAR	0010	06-13-68	SL N	7.0	N 3.00	125		125		101	42	59		
32856	BOAR	0010	02-24-69	SL N	10.0	N 3.00	43		43*		12	12	31		
	LIVESTOCK			*NEW BASE $ 48			**SOLD $	38	*SOLD $		$ 373	$ 153	$ 225		
36001	IH-M TRACTOR	USED	12-01-67	SL N	0.0	N 4.44	$1,000		$1,000	200	$ 820	$ 180	$ 640		
36002	PLOW	NEW	01-01-63	SL N	12.0	Y 10.00	650		650		260	22	238**	05-01	4.0
36003	HOG WATERER	USED	01-01-63	SL N	12.0	Y 10.00	56		56		22	5	17		
36004	DISC	NEW	01-01-64	SL N	12.0	Y 10.00	621		621		311	63	248		
36005	ELEVATOR	NEW	01-01-65	SD N	12.0	Y 10.00	560		560		214	25	189**	06-15	5.0
36006	PUMP	NEW	01-01-65	SD N	12.0	Y 10.00	130		130		50	15	35		
36007	MOWER	NEW	01-01-65	SD N	12.0	Y 10.00	345		345		132	38	94		
36008	STALK CUTTER	USED	01-01-65	SD N	12.0	Y 10.00	396		396		151	43	108		
36009	GMC PICKUP	NEW	11-07-66	D2 N	10.0	N 4.00	1,875		1,875		273	136	137		
36010	AUTO 1/2 OF 2375	NEW	01-08-67	D2 N	7.5	N 4.00	1,187		1,187		886	148	738		
36011	WAGON	USED	09-08-67	SL N	4.5	N 12.00	150		150	25	133	12	121		
36012	JD 3 BTM PLOW	USED	05-01-68	SL N	0.0	N 5.00	550		550	50	513	100	413		
36013	N I STALK SHREDDER	USED	12-12-68	SL N	8.0	N 5.00	100		100	25	100	15	85		
36014	JD BALER EJECTOR	USED	05-01-69	SL N	8.0	Y 7.00	1,000	238	1,238	200	1,238*	99	1,139		
36015	I H HAY CONDITIONER	USED	06-15-69	SL N	7.0	Y 5.00	150	189	334	30	339*	35	304		
36016	N I 506 MAN LOADER	NEW	07-15-69	D2 N	6.0	N 10.00	525		525	100	525*	147	378		
	MACHINERY & EQUIPMENT			*NEW BASE $ 2,102			*SOLD $	427	*SOLD $		$ 3,865	$ 1,083	$ 4,457		
37001	BARN 1/2 INT		01-01-53	SL N	12.0	N 33.33	$ 500		$ 500		$ 260	$ 15	$ 245		
37002	CORN CRIB		01-01-64	SL N	12.0	Y 10.00	980		980		490	98	392		
37003	WATER PUMP & SYSTEM		01-01-64	SL N	12.0	Y 10.00	872		872		436	87	349		
37004	FENCE		01-01-64	SL N	12.0	N 10.00	264		264		132	26	106		
37005	FENCING		01-01-65	SL N	12.0	N 10.00	418		418		251	42	209		
37006	POLE BARN 1/2 INT		01-01-65	SL N	12.0	N 25.00	2,600		2,600		2,184	104	2,080		
37007	SILO		01-01-66	SL N	12.0	N 25.00	4,098		4,098		3,606	164	3,442		
37008	BARN 1/2 INT		01-01-66	SL N	12.0	N 20.00	500		500		425	25	400		
37009	POLE BARN 1/2 INT		01-01-66	SL N	12.0	N 24.00	2,600		2,600		2,275	108	2,167		
37010	FENCE		01-01-66	SL N	12.0	N 10.00	837		837		586	84	502		
37011	SILO NO 2		01-01-69	SL Y	12.0	Y 25.00	4,645		4,645		4,645*	186	4,459		
37012	POLE BARN		01-01-69	SL N	12.0	N 25.00	1,820		1,820		1,820*	73	1,747		
	BUILDINGS & IMPROVEMENTS			*NEW BASE $ 6,465							$10,645	$1,012	$16,098		
	TOTAL SCHEDULE			*NEW BASE $ 8,610			*SOLD $	465			$14,883	$ 2,246	$20,780		

STRAIGHT LINE $ 1,696   DECLINING BALANCE $ 325   SUM OF DIGITS $ 121   EXTRA 1ST YR$ 105
EXTRA FIRST YEAR DEPRECIATION BASED UPON CORPORATE OR JOINT RETURN

Form 9.14. Format of an EDP Net Worth Statement (courtesy Federal Intermediate Credit Bank, Omaha, Nebr.)

AGRIFAX

| 01-01- MODEL 200 | ROBERT J SAMPL
R R 1
ANYTOWN U. S. A. 55551 | MEMBER NO 02-06-01-05432
OWNER NO 11
PAGE   NO 1   06/20/ |

FINANCIAL STATEMENT

ASSETS			LIABILITIES	
11 HOL COW	32	8,700.00	05 PCA	58,188.00
11 BRED HFRS	13	2,600.00	51 FEED	331.00
11 OPEN HFRS	17	2,040.00	08	1,500.00
11 CALVES	24	960.00		
12 STEERS	20	2,400.00		
15 LAYING HENS	6,317	1,895.10		
20 CORN	2,400 BU	2,400.00		
20 OAT	3,000 BU	1,800.00		
20 HAY	3,000 BAL	1,200.00		
20 STRAW	1,500 BAL	450.00		
20 HAYLAGE	150 TON	1,500.00		
20 CORN SILAGE	150 TON	900.00		
20 CONCENTRATE	3 TON	200.00		
51 A/REC		150.00		
52 SUPPLIES		480.00		
01 FARM CHECKBOOK		1,437.97		
71 SAV & OTHER CAS		25.00		
70 PCA STOCK		4,060.00		
98 OTHER ST & BOND		1,800.00		
98 CASH VAL INS		600.00		
28 MACH		35,160.00		
28 AUTO		2,340.00		
TOTAL CUR ASSETS		**$73,098.07**	**TOTAL CUR DEBTS**	**$60,019.00**
70 REAL ESTATE		100,875.00	08 INDIVIDUAL	34,000.00
TOTAL ASSETS		$173,973.07	TOTAL DEBTS	$94,019.00
			NET WORTH	$79,954.07
			CHANGE NET WORTH	$5,821.19

	19 —	19 —
TOTAL CUR ASSETS	74,144.00	73,098.07
TOTAL ASSETS	176,644.00	173,973.07
TOTAL CUR DEBTS	67,511.12	60,019.00
TOTAL DEBTS	102,511.12	94,019.00
NET WORTH	74,132.88	79,954.07

`" I hereby certify that this statement is true and correct.`

Date signed: _____   _____ "
                                              Signature

175

Form 9.15. Total Farm Business Analysis Report (courtesy Iowa Farm Business Assoc., Colo, Iowa)

YEAR 1967   COUNTY 00   FARM NO. 999   ASSOCIATION NO. 99

| | YOUR FARM ANALYSIS | | | 320 ACRE SIZE GROUP | | | |
	OPERATOR	LAND-LORD	TOTAL FARM	HIGH PROFIT THIRD	AVG. PROFIT	LOW PROFIT THIRD	AREA AVG.
MANAGEMENT RETURN							
DISTRIBUTION OF NET FARM INCOME	$ 8340	3558	11898	14711	8012	849	11202
PLUS INTEREST PAID	$ 3534	0	3534	2033	2150	2971	2701
LESS 7 PERCENT INT. ON WORKING ASSETS	$ 3947	0	3947	3552	3263	3159	4688
LESS 5 PERCENT INT. ON FIXED ASSETS	$ 3000	3400	6400	4931	4431	4183	6111
LESS OPERATOR LABOR AT $400 PER MONTH	$ 4800	0	4800	4773	4803	4788	5094
LESS FAMILY LABOR AT $300 PER MONTH	$ 0	0	0	334	304	265	426
** MANAGEMENT RETURN	$$$ 127	158	285	3154	-2639	-8575	-2416
SIZE OF BUSINESS							
FEED AND LIVESTOCK INVENTORY	$ 47386	0	47386	38742	35689	35482	52239
MACHINERY AND EQUIPMENT INVENTORY	$ 9004	0	9004	11854	10800	9525	14547
LAND AND IMPROVEMENTS INVENTORY	$ 60000	68000	128000	98615	88630	83663	122222
TOTAL CAPITAL MANAGED	$$$ 116390	68000	184390	149211	135118	128669	189005
TOTAL FARM ACRES OWNED PLUS RENTED	320		320	317	310	301	419
TOTAL CROP ACRES	275		275	249	235	227	316
TOTAL ROTATED ACRES	290		290	265	255	248	342
MAN-MONTHS OF LABOR	14		14	15	14	15	18
SOURCES OF INCOME GROSS PROFITS							
TOTAL GROSS PROFITS	$$$ 30993		30993	32239	25333	19223	35427
GROSS VALUE OF CROPS PRODUCED	$ 23007		23007	21374	18304	16314	26207
LIVESTOCK INCREASE OVER FEED	$ 7174		7174	9113	5953	2479	7821
MISCELLANEOUS	$ 812		812	1751	1076	430	1399
LAND USE AND CROP PRODUCTION							
CORN GRAIN ACRES	113		113	138	123	114	168
YIELD BU. PER ACRE	96		96	96	88	80	90
CORN SILAGE ACRES	20		20	5	7	10	16
YIELD TONS PER ACRE	15		15	12	12	11	13
OAT ACRES	48		48	12	14	13	13
YIELD BU. PER ACRE	73		73	58	54	47	56
SOYBEAN ACRES	55		55	46	41	32	54
YIELD BU. PER ACRE	27		27	33	28	25	28
HAY ACRES	39		39	19	25	29	27
YIELD TONS PER ACRE				3	3	3	4
OTHER ROW CROPS ACRES	0		0	11	6	5	10
DIVERTED ACRES	0		0	14	18	21	24
TOTAL DIVERTED ACRES PAYMENT	$ 0		0	923	1128	1365	1572
DIVERTED ACRES PAYMENT PER ACRE	$ 0		0	66	64	64	64
ROTATED PASTURE ACRES	15		15	17	20	21	27
PERCENT ROTATED LAND IN ROW CROPS	65		65	76	69	65	72
FERTILIZER COST PER ROTATED ACRE	$ 7.48		7.48	8.63	9.28	10.29	10.86
FERTILIZER COST PER CORN ACRE	$ 16.30		16.30	16.00	18.29	20.53	20.23
** GROSS VALUE OF CROPS PER ROTATED ACRE	$ 79.33		79.33	79.89	70.97	65.02	75.77

Form 9.15. (continued)

	YOUR FARM ANALYSIS			HIGH PROFIT THIRD	AVG. PROFIT	LOW PROFIT THIRD	AREA AVG.
	OPERATOR	LAND-LORD	TOTAL FARM				
LIVESTOCK PRODUCTION							
NET LIVESTOCK INCREASE	$		28664	28101	24246	20804	35953
VALUE OF FEED FED TO LIVESTOCK	$		22490	18988	18293	18325	28132
** LIVESTOCK RETURNS PER $100 FEED FED	$$$		133	148	133	114	128
BEEF INCREASE PER CWT.	$		22.13	23.58	24.13	24.19	24.25
NO. OF CATTLE PURCHASED			149	81	89	92	167
NO. OF CATTLE SOLD			142	89	94	97	175
WEIGHT INCREASE	$		58125	51593	50646	53848	88247
VALUE INCREASE	$		12416	11656	11050	11181	19751
BEEF FEED RETURN	$		121	129	118	103	121
NO. OF BEEF COWS			0	14	13	14	14
HOG INCREASE PER CWT.	$		17.85	18.91	19.54	18.99	19.08
NO. OF HOGS PURCHASED			2	21	31	23	41
NO. OF HOGS SOLD			414	355	311	240	362
WEIGHT INCREASE	$		90790	76723	64634	47083	81464
VALUE INCREASE	$		16248	15342	12061	8577	14557
HOG FEED RETURN	$		145	162	149	117	147
PIGS WEANED PER LITTER			6.75	7.92	7.38	6.73	7.51
TOTAL NO. OF LITTERS			63	50.9	42.7	33.3	46.3
TOTAL PIGS WEANED			425	404	315	225	348
DAIRY INCREASE PER COW	$		0	364	331	322	356
NO. OF DAIRY COWS			0	2.8	2.7	2.6	3.7
VALUE INCREASE	$		0	1023	884	842	1318
DAIRY FEED RETURN	$		0	0.54	0.06	0.00	0.03
POULTRY INCREASE PER UNIT	$		0	2.66	3.97	3.55	3.65
NO. OF HENS			0	32	58	59	55
VALUE INCREASE	$		0	86	230	209	203
POULTRY FEED RETURN	$		0	69	112	84	106
TOTAL OTHER LIVESTOCK INCREASE	$		0	-6	22	-5	125
FEED FED TO OTHER LIVESTOCK	$		0	0	0	0	0
MACHINERY AND LABOR USE							
** MACHINE AND POWER COST PER ROTATED ACRE	$		22.72	22.88	23.19	23.93	24.11
** MACHINE AND POWER INVESTMENT PER ROT. ACRE	$		31.05	44.47	41.98	38.22	44.46
ROTATED ACRES PER MAN			248	216	212	204	228
LIVESTOCK INCREASE PER MAN	$		24499	22876	20127	17171	23935
** GROSS PROFITS PER MAN	$		28490	26245	21029	15867	23585
INCOME AND COST RATIOS							
GROSS PROFITS PER ACRE	$		96.85	101.56	81.84	63.82	84.60
TOTAL EXPENSE PER ACRE	$		59.67	55.22	55.96	61.00	57.85
NET FARM INCOME PER ACRE	$		37.18	46.34	25.88	2.82	26.75
** GROSS PROFITS PER DOLLAR EXPENSE	$		1.62	1.84	1.46	1.05	1.46
NET FARM INCOME PER DOLLAR EXPENSE	$		.62	.84	.46	.05	.46

Form 9.16. Corn-Livestock Enterprise Analysis (courtesy Pioneer Data Systems Inc., Des Moines, Iowa)

PLEASANT VIEW STOCK FARM
ENTERPRISE ANALYSIS-FEED CORN
FOR THE 3 MONTH PERIOD ENDED 06-30-70

	******** SECOND QUARTER ********			******** YEAR-TO-DATE ********		
	DOLLARS	UNITS	UNIT VALUE	DOLLARS	UNITS	UNIT VALUE
REGULAR INCOME						
INV. CHANGE	$ 6,120.00-	6,120.00-	$ 1.00	$ 10,120.00-	10,120.00-	$ 1.00
TOTAL	$ 6,120.00-			$ 10,120.00-		
INTRA-FARM INCOME						
CORN SALES	$ 6,120.00	6,120.00	$ 1.00	$ 10,120.00	10,120.00	$ 1.00
TOTAL	$ 6,120.00			$ 10,120.00		
TOTAL INCOME	$ .00			$ .00		
REGULAR EXPENSE						
SEED PURCHASES	$ 40.00	2.00	$ 20.00	$ 160.00	8.00	$ 20.00
FERTILIZER EX.	$ 118.00			$ 534.00		
INSECTICIDE	$ 80.00			$ 80.00		
TOTAL	$ 238.00			$ 774.00		
INTRA-FARM EXPENSE						
TOTAL	$ .00			$ .00		
TOTAL EXPENSE	$ 238.00			$ 774.00		
TOTAL PROFIT OR LOSS	$ 238.00-			$ 774.00-		
NET WORTH GAIN OR LOSS	$ 6,358.00-			$ 10,894.00-		

178

Form 9.17. Dairy Enterprise Analysis (courtesy Federal Intermediate Credit Bank, Omaha, Nebr.)

AGRIFAX®

12-31-
MONTH ENDING
MODEL 200

ROBERT J SAMPLE
R R 1
ANYTOWN U. S. A.   55551

02-06-01-05432
OWNER NO 11

* FINAL REPORT *   E N T E R P R I S E   A N A L Y S I S   5110   DAIRY

	YEAR TO DATE
AVERAGE NUMBER COWS	34
POUNDS MILK SOLD/COW	15,128
PRICE/CWT MILK	$4.61
OTHER INCOME/CWT MILK	$.60
TOTAL INCOME/CWT MILK	$5.21
MILK INCOME/COW	$698.01
OTHER INCOME/COW	$90.50
TOTAL INCOME/COW	$788.51
COST PURCHASED FEED/CWT MILK	$.24
COST RAISED FEED/CWT MILK	$1.62
PRODUCTION COSTS/CWT MILK	$2.32
TOTAL COSTS/CWT MILK	$2.32
TOTAL COSTS/COW	$350.98
MARGIN/CWT MILK	$2.29
HOURS OPERATOR & FAMILY LABOR	* NOT REPORTED

* HAVE YOU ACCURATELY REPORTED THE INFORMATION REQUESTED

valued, and recorded. Veterinary and medicine expenses must be itemized for each lot. Fixed costs and facilities, including depreciation on equipment items, must be allocated to each lot; and operator labor charges should be specified for each group of hogs. Often EDP subscribers fail to realize the amount of detailed information necessary for accurate enterprise analysis reports.

Historical Summary

Form 9.18 illustrates another capability of the computer. With proper data storage, reports summarizing key information over a period of years can be produced. In addition to the balance sheet in Form 9.18, historical summaries of income statements, financial ratios, and farm production can be provided. Summaries such as these present a concise long-run picture of the farm business operation. This type of information facilitates easy identification of trends and better analysis of what really has been occurring in the farming operation.

EXTERNAL SERVICES PROVIDED

External services most commonly provided as part of EDP packages include farm management consultation, farm tours, farm analysis meetings, tax service, and forward planning. These external services are most often associated with EDP programs offered by farm business associations and universities because these are the organizations most frequently providing full-time area farm management fieldmen. Through periodic farm visits, these men provide personal management consultation for each subscriber. Farm analysis group meetings offer the opportunity for the fieldmen and subscribers to analyze trends in farming, market prices, government farm programs, and other new developments appearing in agriculture. Farm tours offer the opportunity to see efficient methods and ideas in operation. Forward planning services take a variety of forms. The most common examples are projected cash flows, partial budgeting, and linear programming.

GUIDELINES FOR SELECTING AN EDP SYSTEM

This discussion of EDP has covered several methods of processing farm records and accounts on computers as well as types of available packages and basic operational procedures. But even with this background of information, one may still be uncertain about which record-keeping program is best.

Farm Record Needs

The first step is to determine your information needs; it is important to realize that these needs are unique. They depend upon such factors as the size, type, and complexity of your farming operation; your age; your educational background; your training and experience in record keeping and available time; and how the information will be used. The following checklist contains the types of information available in EDP packages. Check the information you need for planning and operating a profitable farm business. It is necessary to know not only the type of information needed but also how frequently it is desired.

Form 9.18. Historical Summary of Income Statements (form John R. Schlender, Ph.D. Dissertation, Purdue Univ., 1970)

John P. Recorder
R. 1
Hometown, Indiana

Item	Balance sheet				
	Dec. 1965	Dec. 1966	Dec. 1967	Dec. 1968	Dec. 1969
ASSETS					
Farm Assets					
Current assets:					
Beef cattle inventory	$ 36,478	$ 37,680	$ 31,231	$ 27,240	$ 26,036
Hog inventory	3,295	3,122	5,675	6,980	8,860
Corn inventory	5,650	7,264	2,984	9,622	7,948
Combined feed inventory	b	b	2,609	b	2,400
All other current assets[a]	460	210	77	180	346
Total Current Assets	45,883	48,276	42,576	44,022	45,589
Fixed assets:					
Land	124,354	124,354	124,354	124,354	125,054
Buildings and improvements	10,680	9,874	15,486	15,142	16,240
Machinery and equipment	9,675	11,240	10,897	11,476	14,568
Total Fixed Assets	144,709	145,468	150,737	150,972	155,862
Total Farm Assets	190,592	193,744	193,313	194,994	201,451
Nonfarm Assets					
Current assets:					
Cash value, life insurance	1,145	1,238	1,342	1,448	1,550
Cash in bank	480	b	1,460	787	650
Bonds	500	500	500	500	500
Accounts receivable	b	b	b	b	468
All other current assets[a]	28	23	41	52	132
Total Current Assets	2,153	1,761	3,343	2,787	3,300
Fixed assets:					
Buildings and improvements	18,468	19,450	21,592	21,376	23,550
Total Fixed Assets	18,468	19,450	21,592	21,376	23,550
Total Nonfarm Assets	20,621	21,211	24,935	24,163	26,850
Total Assets	211,213	214,955	218,248	219,157	228,301
LIABILITIES					
Current liabilities:					
Notes payable	31,956	26,820	25,225	19,146	15,400
Accounts payable	b	475	b	769	600
Taxes payable	b	375	b	501	504
All other current liabilities[a]	160	240	138	342	287
Total Current Liabilities	32,116	27,910	25,363	20,758	16,893
Fixed liabilities:					
Real estate mortgages	95,080	93,880	92,680	91,680	88,480
Notes payable	b	b	4,240	3,270	3,020
All other fixed liabilities[a]	541	274	515	182	460
Total Fixed Liabilities	95,621	94,154	97,435	95,132	91,960
Total Liabilities	127,737	122,064	122,798	115,890	108,751
Net Worth	83,476	92,891	95,450	103,267	119,560
Net Worth + Liabilities	211,213	214,955	218,248	219,157	228,245

[a]Includes items amounting to less than 2% of group total

[b]Included in "all other." Amount is less than 2% of group total.

Information available

_____ Summary of input entries
_____ Ledger account summaries
_____ Cash-flow reports
_____ Income statements
_____ Net worth statements (balance sheets)
_____ Tax work sheets
_____ Crop inventory listing
_____ Livestock inventory listing
_____ Depreciation schedule
_____ Capital purchases and sales
_____ Enterprise analysis reports
_____ Farm business analysis reports
_____ Family living summaries
_____ Accounts payable
_____ Accounts receivable
_____ Summary of principal and interest payments
_____ Payroll summaries

Comparative analysis reports

_____ Actual performance compared with projected plans
_____ Performance this year compared with that of previous years
_____ Your farm compared with other farms of similar resources or production activities

Accounting Methods

One can next evaluate, in terms of the information needed, whether cash, accrual, single-entry, or double-entry recording is most desirable. The information needed will very likely indicate which method to use.

External Services

External services are an integral part of some EDP record-keeping programs and are not offered in others. Some EDP firms do not have the type or number of personnel needed to offer these services. Select a program that provides the type of external services you need to operate an efficient farm business. Again, your needs are unique. Some farm operators do not desire external services, while others have found them to be very helpful. The following checklist contains some of the more common external services offered. Check the services you need and desire in an EDP program.

_____ Management consultation
_____ Farm tours
_____ Tax service
_____ Farm analysis meetings
_____ Forward planning assistance

Operational Procedures

Operational procedures are the heart of any EDP record-keeping pro-gram. It is extremely important that the code system, recording procedures, output reports, and assistance available be carefully evaluated before selecting a program. The code system is used for every recorded transaction; thus it should be simple and easy to use. To receive accurate information, the recording procedures must be followed ex-plicitly. The procedures should be as simple as possible and still permit the type of record keeping you desire. The output reports must be clear and understandable and provide the information needed. Do not select a more elaborate system than you have time to maintain during peak labor periods or one that provides information you do not need. And finally, the type of assistance available can play an extremely large role in your success with EDP record keeping.

The following checklist contains features and points to consider in evaluating EDP record-keeping programs. Check the features contained in the EDP program you are considering.

Operational Procedures

Code system

_____ It is easy to understand and use.

_____ Code categories are appropriate for the record entries of your farming operation.

_____ Additional code categories can be assigned if desired.

_____ It permits both general and detailed levels of record keep-ing.

_____ It permits recording receipts and expenses to individual livestock groups and crop fields for enterprise analysis.

Recording procedure

_____ Input forms have as much uniformity and conciseness as possible (i.e., consistency of column headings, location of code numbers, etc.).

_____ It permits common terminology and quantity units to be used in recording input information.

_____ It provides for itemizing group purchases (e.g., several different items purchased at a farm supply store).

_____ It provides for recording capital purchases and deprecia-ble assets.

_____ It provides for recording intrafarm transfers for enter-prise analysis.

_____ It provides for recording inventory information.

_____ It has subscriber code entries.

_____ It has servicing organization code entries.

_____ Error correction procedure is simple and easy to use.

_____ It permits the degree of detail desired in recording infor-mation.

_____ It permits the method of record keeping desired (i.e., cash, accrual, single-entry, double-entry).

_____ It offers the desired frequency of submitting information to the servicing organization.

Output reports

_____ Reports have as much uniformity as possible (e.g., location of headings, information, etc.).

_____ Terminology is clear and understandable.

_____ They provide the type and amount of information desired.

_____ They provide the degree of detail desired.

_____ They provide information at the desired frequency.

_____ They provide a listing of all transaction information submitted to the computer to aid in locating errors.

_____ The program contains built-in accuracy checks on information submitted to the computer.

_____ The journal summary contains space to write a personal description of each transaction if desired.

_____ Reports can be reconciled with the checkbook balance.

_____ There is an acceptable length of turn-around time (i.e., days elapsed from date of information submittal to date of receiving output reports).

Assistance available

_____ From personal assistance locally

_____ By telephone

_____ By group meetings

_____ From visitation program

_____ When requested only

_____ At an additional fee

_____ Program provides type of assistance and external services desired.

ADVANTAGES AND DISADVANTAGES OF EDP PACKAGES

It is important to recognize the basic advantages and disadvantages of each type of EDP package discussed in preceding sections.

Year-end Farm Summary Programs

In year-end farm summary programs information is recorded at the farm throughout the year and usually no code system is used. The transactions are entered in the proper place in the record book and are totaled at the end of the year for computer processing.

One disadvantage of year-end farm summary programs is that output reports are produced only at the end of the year. The information is available at all times in the form of recorded entries in the record book, but some hand calculations must be performed for it to be useful.

Coded Check Programs

One of the attractive features of coded check programs is that they are simple and easy to use. The only requirement is to add the proper code number to a check or deposit slip. Since this type of program is usually offered by banks, personal assistance is available locally, which is an important advantage. Other advantages include providing monthly output reports, which in some programs are reconciled with the bank statement thus providing an accuracy check on the information. It is advisable to select a coded check program that employs multiple-entry checks to permit itemizing group purchases.

One of the major disadvantages of most coded check programs is that they are limited to basically cash-flow record keeping. However, some coded check programs do have supplemental entry forms for recording inventory information and other non-checkbook type transactions. Also, depreciation schedules are sometimes offered. A problem encountered in some coded check programs is that the procedure for recording refunds to specific code categories may be somewhat confusing.

Mail-in Programs

Code numbers for mail-in programs are characteristically of more digits than those of coded check programs. This permits more detailed record keeping and allows mail-in programs to accommodate accrual recording much easier than coded check programs. As a result, mail-in programs offer more flexibility in the type and amount of information provided. By requiring monthly or quarterly submittal of input information, mail-in programs can act as a stimulus for keeping farm records up to date. However, one should not mistake this as a guarantee of up-to-date farm records.

Since the servicing organizations for mail-in programs may be quite distant, personal assistance may not be as readily available as with coded check programs at a local bank. Assistance is more likely to be provided by telephone or by area group meetings.

Costs

Costs are also important to consider in evaluating EDP programs. Depending upon the size of the farming operation and the specific EDP program selected, the cost of the services can vary from $50 to $1,000 per year, although the typical cost is $100 to $200. Some EDP programs involve a flat fee for the basic program and additional charges for the optional reports. Other firms base their fees on the level of gross farm income. A third method is to charge a small fee on a per transaction basis. Regardless of the method used to assess fees, one should remember that the cost of the program is not always a measure of its quality.

Personal Background

A final consideration in selecting an EDP program is your own background in record keeping. If you have had limited training or experience, it may be advisable to start with a fairly simple EDP program or one that permits flexibility in the level of participation. Over time as experience is gained, you may advance to programs offering more detailed information.

When subscribing to an EDP program, you must be willing to learn new record-keeping procedures. These can be a source of confusion and error until they are mastered fully. New terminology will be encountered, both in entering information on the input forms and in receiving the output reports. You must be willing to undertake these new procedures and responsibilities to receive the benefits offered through EDP.

The computer is a machine that can be used to help you keep accurate farm records. Through EDP some of the pencil pushing in record keeping can be computerized, but subscribing to an EDP program is not a guarantee that you will have accurate farm records. No matter how fancy and professional the printout reports may look, the output information is only as good as the information you put into the machine.

EDP IN THE FUTURE

In the future there will probably be more variation in the type of services offered as part of the EDP program. Some firms will keep their programs simple and offer a minimum amount of information and services in conjunction. Other organizations will expand their services to offer a more complete package to meet the needs of the farmer not only in record keeping but in the area of tax service as well. Providing more management information and forward planning tools are goals of several organizations offering EDP services.

Using the computer to summarize farm records is only scratching the surface of the capabilities of present-day computers. Until the proper software (computer programs) is developed to explore new areas in EDP, much of the potential of the computer in farm management will remain untouched.

10

INCOME TAX MANAGEMENT

TAX MANAGEMENT—PART OF FARM MANAGEMENT

Managing a modern farm business requires the investment of a large amount of capital and the handling of large sums of money annually. The tax consequences of farm business decisions have a greater impact on cash flow and net income as farm businesses become larger.

The farm manager is constantly making decisions during the year that affect the amount of income tax to be paid and the amount of cash available for operation of the business. To make wise decisions in the framework of minimizing income tax while maximizing after-tax income, he must understand the tax consequences of various farm business transactions. Thus, farmers must think taxes throughout the year, as tax management is a continuous process, not just a year-end endeavor.

Tax management assumes that the individual farmer can do a better job by planning his investments and financing and managing his income and expenses than he can by chance or unplanned business operations. The timing of transactions can play an important role in the short run in balancing income between years and thus minimizing yearly fluctuation in income and taxes. An attempt should be made to maintain an annual net income at least equal to the year's allowable nonbusiness deductions and personal exemptions, and yet avoid extremely high taxable income. It is equally desirable to have a net income that approaches or equals the maximum earnings (from self-employment plus wages) eligible for social security credits. Tax management aims at the greatest after-tax income and net worth.

In the long run, however, tax management involves much more than merely timing transactions between two years to even out income. The form of business organization under which the farm business is operated and the manner in which enterprises are organized within the farm business as well as strategies employed in financing the business have a long run effect on after-tax income.

Tax management thus is not concerned solely with minimizing taxes. If decisions are made and business transacted solely in an effort to reduce tax, net income *after* taxes

---

This chapter is adapted from Income Tax Management for Farmers, North Central Regional Publ. 2, 1975.

may actually be lower. For example, if a decision results in the saving of $100 in income tax, but a larger amount is lost by a lower selling price of a farm product, the net income after taxes is reduced. Frequently, there is no conflict between a wise tax decision and a good farm business decision, but when a choice must be made, the one resulting in the larger net income after taxes should be followed.

Income tax regulations often change from year to year. From a tax management consideration the farmer should be aware of changes that may influence his management decisions.

The "Farmer's Tax Guide," published annually by the Internal Revenue Service (IRS), is most helpful in keeping up to date on the frequent changes in income tax laws and in preparing your tax return. It contains current tax information, examples, and illustrations to guide the farm business manager in preparing his income tax return. Get a new copy each year from the IRS or any county extension office.

GOOD RECORDS—A NECESSITY

Successful tax management depends upon keeping complete records regularly and carefully throughout the year. Such records enable the taxpayer to determine approximate taxable income at any time during the year. These records provide a basis for making business decisions that increase net income after taxes.

If a preliminary check of your expected total income and expenses indicates an unusually high taxable farm income, the decision may be made to delay additional sales until after the end of the year or to increase deductible expenditures before the end of the year. Conversely, if the check reveals an unusually low net farm income, sales may be speeded up to include them in the current year, or expenditures or payment of accounts may be deferred until the next year. Such procedures are more advantageous to farmers using the cash method of accounting, since farmers using the accrual method must include all unsold products and unused purchased supplies in the inventory at the end of the year.

Depreciation of farm improvements, machinery, and equipment and purchased breeding, dairy, sporting, and work animals are allowable business deductions under both cash and accrual methods. Farm records of such property should include a detailed record of: date of purchase, cost, years of life, method of depreciation, estimated salvage value, depreciation claimed to date, and remaining cost. Investment credit has been repealed and thus is no longer allowed. However, records of previous investment credit taken should be retained because of possible recapture.

Bank statements are often examined when tax returns are audited. Thus, deposits shown on these statements should clearly show the source of income deposited. Nontaxable income, which should by all means be identified on bank statements, includes borrowed money, amounts received as repayment of loans, bonds cashed, gifts, or inheritances.

Account books suitable for keeping adequate records for tax purposes and farm business analysis as well as "Farmer's Tax Guides" are available at all county extension offices. Machine record systems suitable for income tax and farm business analysis are available from the extension services of several states and from private and cooperative business firms.

RECORDS AND TAX ACCOUNTING METHODS

Farmers may keep records and file their tax returns using either the cash or accrual basis. A combination of methods may be acceptable if it clearly reflects income and expenses. The method of accounting for tax purposes must be the same as the one by which the farmer keeps records.

A farmer makes his choice of filing his tax return on the cash or accrual basis when he files his first tax return. If the farmer is a part of a newly organized farm partnership or corporation or he files for a newly purchased farm operated as a separate business, he may file on the same basis or may change to the other. Having chosen one method he must continue using that method unless written consent to change is obtained from the Commissioner of the Internal Revenue Service, Washington, D.C.

Farmers may report farm business income on the cash basis and other business or personal income on the accrual basis, or vice versa.

Cash Method

When the income tax return is filed on the cash basis, gross farm income includes: income actually or constructively received from the sale of all crops and market livestock produced on the farm, profits (selling price less cost) on livestock and other items purchased for resale, and other farm income actually or constructively received. Allowable deductions include: farm operating expenses paid during the year, regardless of when they were incurred and depreciation expense allowable on farm improvements; machinery; equipment; and purchased dairy, breeding, sporting, and work stock.

There are certain advantages of filing on the cash basis in tax planning:

1. The cash method provides a simple method of reporting. Fewer records are necessary, as inventory accounts need not be kept. However, the inventory information may be needed for business analysis or for preparing a financial statement.
2. Taxes are postponed for the cash-basis farmer who is in a period of year-to-year increases in inventory. He then has a tax advantage, particularly if tax rates remain constant or decrease. The advantage is less if tax rate increases are in prospect as his farm business moves toward maturity and cash sales of products accumulated during years of lower taxes materialize. However, if decreases in tax rates are expected, the opportunity for tax savings increases as products accumulated during years of higher tax rates are sold in years of lower tax rates.
3. More flexibility is provided in the cash method to adjust income from year to year when wide variations may occur in prices and production rates, despite the tendency of the accrual method to automatically make these types of adjustments each year through beginning-of-year and end-of-year inventories.

   For example, when using the cash method, part of the crop or livestock production in a good year may be held over for sale in years of lower production or sales may be speeded up in years of low production and/or prices. One may also delay expenditures, postpone payments, or conversely make certain cash purchases before actually needed, depending upon the net income situation in a particular year.

   Deductible ordinary expense items such as feed, seed, fertilizer, and repairs may be purchased in the latter part of the year, even though they are not used until the next taxable year. However, in making these purchases, care must be exercised. It is usually good management for a livestock feeder to purchase needed feed grain in late fall when prices are generally lower than in the following spring months. Advantages would depend on prices, cost of storage, and cost of money.

Fertilizer, feed, chemicals, petroleum, or other annual operating supplies purchased should be actually acquired and preferably delivered to the farm. They must be bona fide irrevocable obligations, not advance payments on "orders to be placed" or on purchases that may be made in the succeeding year. Expenditures will not be allowed until the obligation is paid.

4. Sales of raised breeding, dairy, sporting, and work stock, treated as capital assets, result in a lower tax liability if the cash method is used because these animals have a zero cost basis when sold, while under the accrual method the cost basis for determining gain is the last inventory value if left in inventory (or salvage value plus remaining depreciation if capitalized). In addition, if animals are capitalized under the accrual method, any depreciation taken is now recaptured as ordinary income if the sale income is greater than the depreciated value.

5. Upon the death of a farmer who has filed tax using the cash method, the unsold livestock, crops, and other farm commodities pass to his estate without being taxed as income. The estate takes the property with a tax basis equal to the fair market value at death. For a farmer on the accrual method, the values of these items are included in the inventory of his final return.

Accrual Method

When the tax return is filed on the accrual method, gross farm income includes: all income from sales made during the year regardless of when payment is received, all miscellaneous income regardless of source, and the inventory value of all livestock, crops, and supplies on hand not sold at the end of the year.

To arrive at net income, the farmer subtracts: the inventory value of livestock and products on hand at the beginning of the year, the cost of livestock or products purchased during the year (except livestock held for draft, breeding, dairy, or sporting purposes unless they are included in inventory), all operating costs or expenses incurred during the taxable year, and depreciation—the same as allowable under the cash method.

Costs of purchased feeder livestock are subtracted in the year purchased and then will be included in the inventory at the end of the year if not yet sold. With the cash method these costs are not deductible until the year in which the animals are sold. (See also the discussion of accounting methods in the "Farmer's Tax Guide.")

Some advantages of the accrual method include:

1. Farmers, who store some crops and sell them in the next year, level their income to some extent by including the production of a given year in the closing inventory of the year in which most of the costs of production were incurred. They may thus avoid having to pay income tax in one year on the sales of more than one year's production. An example would be two calf crops or some grain held over plus the current year's production.

Livestock feeders who have heavy expenses in the latter part of one year but sell the livestock early in the following year may prefer the accrual method in order to count the increased value of the animals in their ending inventory and offset costs that may already be paid. Farmers who report on the accrual method have their income tax paid more up to date than do farmers who use the cash method and deduct all expenses but hold back unsold production from one year to the next. The accrual method generally results in a more even year-to-year taxable income.

2. Young farmers starting with a small operation or with inadequate financing may desire to use the accrual method in order to keep their income and taxes on a more current basis, rather than postponing taxes until the year in which production is sold. They may need to count the increase in the values or unsold production as income to offset cash farm operating expenses, allowable nonbusiness deductions, and personal exemptions or to avoid showing a net operating loss.

3. Social security benefits may often be augmented by using the accrual system. The application of Section 1231 to sales of raised breeding, dairy, sporting, and work stock excludes proceeds from such sales from the self-employment income under the cash method. Further, all costs of raising Section 1231 livestock are deducted from self-employment income. But with the accrual method the requirement of capitalization of "normal costs" of raising these animals in the year of sale tend to increase self-employment income.

   The requirement of capitalization of "normal costs" of raising Section 1231 animals in the year of sale may be a disadvantage of the accrual system. With more ordinary income from the sale of breeding stock, hog farmers, dairy farmers, and beef cow herd operators will pay more taxes under the accrual method.

Comparison of Methods

A significant difference between the cash and accrual methods of computing income is illustrated in the following example. A farmer purchases and takes delivery of feed for $500 on December 15 and charges the purchase. He is billed for it January 2 and pays the bill in January. With the cash method, it is a farm expense in January when the bill is paid. With the accrual method, the $500 expense is a deduction in December when the obligation to pay was incurred or "accrued." Any feed on hand at the end of the year would be included in the ending inventory and the $500 debt in accounts payable. Income is similarly treated.

Increases in inventories are included in the income of the accrual method farmer. For example, a farmer raises and feeds livestock during the year but does not sell any. With the cash method he has no income until the livestock are actually sold. With the accrual method he has income in the amount of any increase in the value of livestock and crops on hand at the end of the year compared to the value at the beginning of the year.

Table 10.1 shows the farm income subject to federal income tax under the cash and accrual methods of reporting for different farm situations. These farm situations are summarized below. In all situations except I, it is assumed that operating and fixed expenses are the same for the two methods. The purpose in A, B, and C is to illustrate what happens when a farm-produced product such as grain is added to, or reduced in, inventory. Situations A, D, and E illustrate the changes brought about by purchased feeder livestock. Situations F, G, and H are added to illustrate what happens when breeding livestock that are subject to capital gains are involved. In I the assumption of equal operating costs is relaxed to allow end-of-year purchases to reduce taxes.

TABLE 10.1. Comparisons of net taxable income when determined by cash and accrual methods under different receipt and expense situations

Transaction	A Cash	A Accrual	B Cash	B Accrual	C Cash	C Accrual	D Cash	D Accrual	E Cash	E Accrual	F Cash	F Accrual	G Cash	G Accrual
Receipts														
Misc. receipts	$ 5,000	$ 5,000	$ 5,000	$ 5,000	$ 5,000	$ 5,000	$ 5,000	$ 5,000	$ 5,000	$ 5,000	$ 5,000	$ 5,000	$ 5,000	$ 5,000
Crops	10,000	10,000	11,000	11,000	9,000	9,000	10,000	10,000	10,000	10,000	10,000	10,000	10,000	10,000
Feeders purchased	10,000	10,000	10,000	10,000	10,000	10,000	10,000	10,000	10,000	10,000	0	0	0	0
Feeders raised	0	0	0	0	0	0	0	0	0	0	10,000	10,000	11,000	11,000
Breeding livestock (capital gains)	0	0	0	0	0	0	0	0	0	0	500	250	500	250
Ending inventory:														
Crops and feed	X	5,000	X	4,000	X	6,000	X	5,000	X	5,000	X	5,000	X	5,000
Feeders purchased	X	6,000	X	6,000	X	6,000	X	6,000	X	7,000	X	0	X	0
Feeders raised	X	0	X	0	X	0	X	0	X	0	X	5,000	X	4,000
Breeding livestock	X	0	X	0	X	0	X	0	X	0	X	2,000	X	2,000
Total Credits	$25,000	$36,000	$26,000	$36,000	$24,000	$36,000	$25,000	$36,000	$25,000	$37,000	$25,500	$37,250	$26,500	$37,250
Expenses														
Operating	$10,000	$10,000	$10,000	$10,000	$10,000	$10,000	$10,000	$10,000	$10,000	$10,000	$10,000	$10,000	$10,000	$10,000
Fixed	1,000	1,000	1,000	1,000	1,000	1,000	1,000	1,000	1,000	1,000	1,000	1,000	1,000	1,000
Depreciation	1,000	1,000	1,000	1,000	1,000	1,000	1,000	1,000	1,000	1,000	1,000	1,000	1,000	1,000
Livestock purchased	5,000	5,000	5,000	5,000	5,000	5,000	6,000	5,000	5,000	5,000	0	0	0	0
Beginning inventory:														
Crops and feed	X	5,000	X	5,000	X	5,000	X	5,000	X	5,000	X	5,000	X	5,000
Feeders purchased	X	6,000	X	6,000	X	6,000	X	7,000	X	6,000	X	0	X	0
Feeders raised	X	0	X	0	X	0	X	0	X	0	X	5,000	X	5,000
Breeding livestock	X	0	X	0	X	0	X	0	X	0	X	1,500	X	1,500
Total Debits	$17,000	$28,000	$17,000	$28,000	$17,000	$28,000	$18,000	$29,000	$17,000	$28,000	$12,000	$23,500	$12,000	$23,500
Net Taxable Income	$ 8,000	$ 8,000	$ 9,000	$ 8,000	$ 7,000	$ 8,000	$ 7,000	$ 7,000	$ 8,000	$ 9,000	$13,500	$13,750	$14,500	$13,750
Difference	none		$ 1,000		$ 1,000		none		$ 1,000		$ 250		$ 750	

H. Cash $12,500; accrual $13,750
1,250 (1,000 + 250)

I. Cash $7,500; accrual $8,000
(500)

($1,000 − $250)

Situations H and I are not illustrated, but the reasons for the differences should be apparent after reviewing A-G.

A. Crop-livestock feeding farm. All livestock sold were purchased. All crop and livestock increases are sold; i.e., beginning and ending inventories remain unchanged. The cost of livestock purchased remains unchanged from the year before.

B. Same as A except $1,000 more crops were sold, thus reducing ending crop and feed inventory by $1,000.

C. Same as A except $1,000 of the year's crop is added to the ending inventory, thus $1,000 less crops are sold.

D. Same as A except that the cost of feeders is not the same as in the previous year. Cost of feeders purchased this year is less; i.e., $5,000 compared to $6,000 last year. The closing inventory difference from the purchase price is $1,000, the same as A.

E. Same as A except that the closing inventory value this year is $2,000 over the purchase price as opposed to $1,000 last year. The cost of the feeders was the same.

F. Crop-livestock raising farm. All crop and livestock increase is sold; i.e., beginning and ending inventories remain unchanged. Normal replacements of breeding stock are sold. These have a beginning inventory value of $500 and are sold for $1,000.

G. Same as F except that inventories of feeders raised are reduced through sale by $1,000.

H. Same as F except that the breeding herd is increased by $1,000, thus reducing the sale of feeders by $1,000.

I. Same as A except that $500 of feed is stockpiled (added to inventory) at the end of this year. It should be recognized that next year the advantage will be reversed.

It also should be observed that a difference in one year, except for capital gains, favoring one or the other of the methods may be reversed the following year. The important things to observe are the flexibilities and rigidities of each system. The level of taxable income over time will be the same under the two systems except for capital gains and the effect of increased inventories. However, the amount of tax paid may be somewhat different through income averaging.

Neither of these methods may provide the route to attaining all goals. However, if a farmer assures himself that he will be faced with increasing inventories and decreasing tax rate, the cash method would reduce taxes. With decreasing inventories and increasing tax rate it is apparent that the accrual method would be desirable.

With other combinations of inventory change and rates of taxation the choice is not as clear-cut. For example, with increasing inventories and rates of taxation, the individual using the cash method avoids paying taxes on this year's increase in inventory but also loses out on paying on the last year's accumulation of property at last year's lower tax rate. Similarly, when the accrual method with decreasing inventories and decreasing tax rate is used, the decreases in inventory operate to reduce taxable income, but taxes are paid on the basis of this year's higher rate. However, compared to the cash method, taxable income and thus taxes would be reduced.

Use of the method that will delay or postpone taxes while taking advantage of all personal deductions and exemptions, even with rising rates of taxation, will generally operate to the advantage of most taxpayers. Postponing taxes is discussed later.

MAXIMIZING AFTER-TAX INCOME

The goal of most commercial farmers is to maximize after-tax income.

An individual's income is related to the resources he has and how he combines them to do business. The form of business organization he chooses to operate within can have an effect on taxes paid. How he organizes within an enterprise, particularly some livestock enterprises, will affect the income that is taxable.

For some, reduction of taxes may be accomplished by income averaging. For others, timing transactions to equalize income between years, particularly to level allowable deductions in each tax year will reduce the total tax bill. Additional first-year depreciation and accelerated methods of depreciation used judiciously can assist in leveling income from year to year.

The above actions deal first with maximizing after-tax income through the type of income produced and the manner in which it is taxed and also with minimizing taxes by timing of income and deductions within the former framework. The dual action of both is to maximize after-tax income.

Some actions a manager may take will not necessarily reduce taxes but will tend to postpone them. Use of additional first-year depreciation and accelerated methods of depreciation, unless used to even income, tend to postpone taxes. The same is true if soil and water conservation expenses and land clearing expenses are charged off as current expenses and disposed of within the time period that the new recapture provision allows. However, if the real estate is held for at least ten years, it is possible to convert ordinary income to capital gains income and permanently reduce tax liabilities.

Form of Business Organization

The farm business may be operated as a sole proprietorship, partnership, corporation, or "hybrid form"—a corporation which is generally treated as a partnership for income tax purposes. With graduated tax rates on individual income and flat rates at two levels for corporate income, taxes paid as a result of the operation of the farm business may differ considerably over a period of years depending upon which form of business organization is used. This will be particularly true for larger taxable incomes. However, generally a farm operator should choose the form of business organization that best fits his operational and estate planning objectives rather than solely income tax objectives. Tax considerations may, however, be a motivating factor in selecting a particular type of farm business organization. (See North Central Regional Publication No. 11, "The Farm Corporation.")

Enterprise Organization

The Tax Reform Act of 1969 extended the holding period for horses for draft and breeding purposes and for cattle for breeding or dairy purposes from one year or more to two years or more to qualify for capital gains treatment for animals acquired after December 31, 1969. In addition, the act provides for recapture of depreciation taken on livestock, i.e., that gain on sale of livestock be treated as ordinary income up to the full value of depreciation deducted in previous years. The provision applies to all years after 1969 but only to depreciation taken after 1969.

These provisions have little effect on the farmer with a dairy herd or beef breeding herd who primarily has raised animals. There is still opportunity, for example, for the dairy farmer to save all his raised heifers for the milking herd and bring them into the milking herd before culling, thus converting ordinary income into long-term capital gain when culling and selling the heifers. Placing all raised heifers into the milking herd before culling versus prior culling and retaining just enough for normal replacement rates in a 40-cow dairy herd could produce an additional $1,000 or more of excludable long-term capital gain. Following such a procedure could increase ordinary income from the dairy herd and possibly could upgrade performance of the herd more rapidly.

Income Averaging

Rules governing income averaging have been changed to enlarge the group of eligible taxpayers. Thus farmers who have had substantial increases in income in the last couple of years may find it advantageous to elect to use the income averaging provision.

A person whose averageable income for 1970 or any following year exceeds his average income for the four preceding years by $3,000 may elect to use averaging. Averageable income is the amount by which taxable income for the computation year exceeds 120% of the average base period income. Long-term capital gains and income from wagering and from gifts are now included in averageable income.

A taxpayer who uses income averaging, however, cannot use the alternative method for computing his capital gains or the new maximum rate on earned income.

MANAGEMENT OF INCOME TO REDUCE INCOME FLUCTUATION

New Low-Income Allowance

For 1975, the low-income allowance was $1,600 for single persons ($950 for a married person filing a separate return), the personal exemption $750, and the standard deduction 16% with a $2,300 ceiling for single persons and $2,600 for joint returns. Incomes at or below the low-income allowances shown in Table 10.2 are not subject to federal income tax.

When a preliminary check of income indicates a probable net taxable income less than the amount allowed by the low-income allowance or the standard deduction and personal exemptions, consideration should be given to at least increasing income to the amount of the total allowance, deductions, and exemptions.

Since the low-income allowance or personal exemption and deduction are allowed annually, exemptions not absorbed by current income are automatically lost. Unused exemption credits cannot be carried forward and applied against income of another year. The following example illustrates this principle. The John and Mary Jones and Jim and Jan Smith families each have two children. Their financial situations are shown in Table 10.3.

TABLE 10.2.   Maximum incomes, not subject to income tax, by number of exemptions, 1975

Number of exemptions	1975
Single individual, under age 65	$2,350
Married couple, under age 65	3,400
Family of four, under age 65	4,900
Single individual, 65 or older	3,100
Married couple, 1 spouse 65 or older	4,150
Married couple, both 65 or older	4,900

TABLE 10.3. Incomes and taxes for two families

| | Net income | | | |
Family	First year	Second year	Average income	2-Year tax
Jones	$ 0	$9,800	$4,900	$734*
Smith	4,900	4,900	4,900	0†

*Using 1975 standard deductions, tax rates and personal exemption tax credit (4 x $30).
†Using low-income allowance for 1975.

The Joneses paid income tax but the Smiths did not even though they had the same total net income for the two years. In the first year Jones paid no tax, but could not use the $4,300 earnings that tax regulations permitted before payment of income tax. In the second year the Jones income exceeded exemptions so they had to pay taxes.

Cash Method

If the Joneses used the cash method of reporting, they could have increased their income for the first year by doing one or more of the following:

1. They could have made additional sales before the end of the year. Livestock feeders may have livestock about ready for market that can be sold either in December or January. Sales of some of these animals in December may be desirable from a tax standpoint, even though they bring less gross income. Any optional increase in income in a given year must be weighed against the tax benefits. Other possibilities include selling grain or other farm products, culling dairy and breeding stock or poultry flocks, collecting money due for labor or custom work done, doing additional off-farm work, or selling capital items not needed in the business and on which a gain can be realized.
2. If they were eligible to borrow from the Commodity Credit Corporation (CCC), they might have secured a loan on some of the grains before the end of the year and elected to treat the amount of the loan as income. *Caution:* Once made, the decision to treat CCC loans as income in the year received must be elected in future years.
3. Some expenditures might be delayed and payment of operating expense charge accounts postponed until the next year. In many cases, farmers charge repairs, fuel, feed for livestock, fertilizer, and farm supplies when purchased. Arrangements may be made for carrying such accounts beyond the end of the taxable year. In some states it is optional to delay payment of half of property taxes until the following year, without interest or penalty.

Accrual Method

If John and Mary Jones used the accrual method, opportunities for increasing income late in the year would be more limited. Possible ways of increasing net income in the first year would include:

1. Doing off-farm or custom work for cash wages.
2. Buying feeder stock or other inventory property that would increase in value by the end of the year.
3. Incurring as few deductible expenditures as possible during the balance of the year.

4. Not taking the additional first-year depreciation allowance on eligible items and not electing to use accelerated methods of depreciation on assets purchased during the year.

Planning Farm Operating Expenditures

Some expenditures may not be made every year but are nevertheless deductible in the year in which they occur. Such expenditures include painting buildings, minor repairs on improvements, many small shop tools, periodic seeding of legumes and grasses, and (within limits) costs of soil and water conservation. Farmers can manage to make many expenditures of this type in years of high gross income to reduce their taxable income.

DEPRECIATION AND DEPRECIATION METHODS

Depreciation is the amount or portion of the cost of a capital asset that is used up or disappears with use or age. (Additional information on depreciation and depreciation methods is contained in Chapter 2.) For income tax purposes it is expressed in dollars and includes obsolescence as well as normal "wear and tear" due to use. By claiming depreciation as a farm business expense, a farmer recovers the costs of farm improvements; machinery; equipment; and purchased breeding, dairy, and work stock during the years of their useful life in his business. Farmers should claim all allowable depreciation since it reduces their taxable income.

There is little advantage, however, in depreciating an item below its estimated market value since the recovery of depreciation (by cash sale) is treated as ordinary income rather than as capital gains. Thus, if excessive depreciation is claimed, it will be restored to income in full if it is "recaptured" by a sale. There is only a temporary advantage in claiming depreciation below a reasonable salvage value.

For good tax management a farmer should realize the impact that the different methods of depreciation and additional first-year depreciation allowance have on taxable income. He can then select the method and options that will provide the best opportunity for reducing taxes and maximizing income in his situation.

Faster depreciation may help the farmer who expects to retire before the useful life of the item expended. It may also be of advantage to young farmers in keeping their tax bill lower and leaving more cash available for debt retirement and business expansion.

On the other hand, if a farmer is expanding his business or expects higher income in future years, he may prefer to continue using the straight-line method on newly purchased items in order to have more depreciation in future years.

LAND DEVELOPMENT COSTS

Farmers are allowed the current deduction of certain soil and water conservation expenses and land clearing expenditures. Soil and water conservation expenses may, if the farmer elects, be treated as currently deductible expenses up to 25% of the gross income from farming in any year. Also, at his election, a farmer may annually deduct expenditures up to $5,000 or 25% of taxable income from farming, whichever is less, for clearing land to make it suitable for farming. This decision (versus capitalization of these costs) defers payment of taxes. It may also reduce the amount of taxes to be paid by converting ordinary income into long-term capital gain if the farm is sold.

The Tax Reform Act of 1969 reduced the advantages of claiming these expenses as current deductions in some instances. The new law provides at the time of sale for the recapture of soil and water conservation and land clearing expenditures made on farmland owned less than ten years. Until it exceeds the value of deductions allowed for such expenditures made after December 31, 1969, gain on the sale of farmland is treated as ordinary income rather than as capital gain. However, there is no recapture if the land has been owned for at least ten years. Expenditures are recaptured in full when land is sold within five years of acquisition. For sales in the sixth through the ninth year, the amount recaptured decreases by 20% each year. The provision is effective for expenditures allowed after December 31, 1969, with respect to farmland sold after December 31, 1969.

Therefore, for most farmers, the opportunity to defer paying taxes and to reduce taxes through converting ordinary income into capital gain remains. Only if the farm is sold within the ten-year period will recapture be required.

DEFERRING TAXES

There are many opportunities within current tax regulations to defer payment of taxes as well as to reduce taxes. These are most often considered in the context of timing transactions and the use of various provisions to avoid fluctuation in income, thus saving taxes through avoiding jumps from one progressively higher rate of taxation to another or avoiding loss of allowable deductions or exemptions.

However, the opportunity to defer taxes for those taxpayers whose income does not fluctuate widely is a valuable consideration. Deferment of taxes from the current year permits more money for immediate family living, permits more rapid increase in net worth, provides additional money for expanding the farm business, and enlarges the individual's borrowing base by providing more internal capital to attract external capital in financing the business. This is particularly valuable to farmers expanding their business.

INVESTMENT CREDIT

The Tax Reduction Act of 1975 provides that for eligible property acquired and placed in service after January 21, 1975, and before January 1, 1977, investment credit is increased from 7 percent to 10 percent.

To qualify, property must be depreciable, have a useful life of at least three years, be tangible personal property or other tangible property, and be placed in service by you during the year. Generally, if property is depreciable (except for buildings), it is eligible for investment credit. Livestock (except horses), machinery, and equipment (except if part of the building) are eligible. For property to qualify under "other tangible property," the property must be used as an integral part of production, etc., or constitute a research or storage facility.

If the useful life of the property is seven years or more, all the investment qualifies; if five to seven years' useful life, ⅔ qualifies; and if useful life is three to five years, ⅓ of the investment qualifies. Under three years useful life, there is no credit.

The above includes only a few of the rules surrounding investment credit. The reader should consult the "Farmer's Tax Guide" and other IRS publications. From a tax management standpoint, the important thing to remember is that investment credit reduces a person's tax bill by the amount of the credit. He should be aware of this when planning eligible expenditures so that he may reduce his total tax bill over time.

TAX REFORM ACT OF 1969

Several changes made by the Tax Reform Act of 1969 have been discussed, but others also have implications for many farmers.

Crop Insurance Proceeds

Changes in handling crop insurance proceeds should help to balance income for cash-method farmers. Originally, crop insurance proceeds were reported as income in the year received. Thus farmers who received insurance proceeds as a result of crop destruction or damage and who customarily held crops harvested in one year for sale in the following year were required, in effect, to report income from two crops in one year.

Under the new law, farmers may now elect to defer reporting these proceeds for federal income tax purposes until the year following the year of loss. It must be established that, under usual practice, income from the damaged or destroyed crops would have been reported in the year following the year of the loss. This provision was made effective for 1969 and subsequent years.

Real Estate Depreciation

The Tax Reform Act of 1969 limits the so-called fast methods of depreciation on new Section 1250 property (depreciable real property, primarily buildings and their components). For such property acquired after July 24, 1969, only the 150% declining-balance depreciation method may be used. The 200% declining-balance and sum-of-the-years-digits methods of real estate depreciation are now limited to new rental residential housing.

In the case of used buildings (except used residential rental housing), depreciation on acquisitions made after July 24, 1969, is limited to the straight-line method.

On the sale of Section 1250 property (other than residential property) after December 31, 1969, accelerated depreciation taken after December 31, 1969, in excess of allowable straight-line depreciation, is to be recaptured as ordinary income to the full extent of the gain.

Capital Losses of Individuals

Previously, an individual could deduct net long-term capital losses from ordinary income to the extent the losses exceeded capital gain, up to a $1,000 limit. Now only 50% of an individual's net long-term capital losses may be used to offset ordinary income up to the $1,000 limit. Thus $2,000 of long-term capital losses are required to obtain the $1,000 maximum deduction. In the case of married persons filing separate returns, the deduction from ordinary income is limited to $500 for each spouse. Formerly, each spouse could claim the $1,000 limit.

Long-term capital losses can be deducted, dollar for dollar, from long-term capital gains, and any excess of such losses over long-term gains can be deducted, dollar for dollar, from net short-term capital gains. The full value of short-term capital losses can be deducted from net short-term capital gains and ordinary income. Also, long-term capital losses in excess of the amount deductible can still be carried over for use in a subsequent year.

Involuntary Conversions

The new law extends the time available for reinvestment of the proceeds from involuntarily converted property in similar use or related use property
from one year after the year in which the involuntary conversion occurs (old provision) to
two years (new provision). The impact of this provision is somewhat muted since one-
year extensions for reinvestment were frequently granted by the IRS under the old law.

NET OPERATING LOSSES

Farmers often pay more taxes over a period of years than required by
law because they fail to take advantage of net operating loss provisions. If a farmer
has a net operating loss in a given year, it can be used to reduce net farm income of other
years. The loss must first be carried back three years and applied against taxable income
of that year. If that taxable income was not sufficient to offset the operating loss, the
remaining excess of the loss is carried to and applied against the income of the second
preceding year and then to the immediate past year. If there is still a remaining excess of
loss over the total taxable income of the three prior years, it is then carried forward to
each of the next five years, or until all is used to offset income.

When a net operating loss occurs, a claim for refund must be filed to recover any tax
paid in prior years and to establish the amount of loss, if any, to be carried forward to
offset future income. Some adjustment of taxable income of prior years may be required.
A qualified tax consultant or an IRS agent should be consulted. If the loss is substantial,
this opportunity to recover income tax paid in prior years should be investigated and not
overlooked. The claim can be filed at any time within three years after the return for the
year of loss was filed. Special tax forms 1045 or 843 are used for this purpose.

SALES OR TRADES OF PROPERTY

When a farm is sold or traded for another farm or business property, it
is necessary to establish the cost basis of the farm sold in order to compute the
actual gain or loss on the transaction and to ensure that no unnecessary tax is paid. In
order to establish this cost basis. it is necessary to have a record of original cost, costs of
all improvements made on the property since it was acquired, and all depreciation
claimed. Some states have special depreciation and investment record books available at
county extension offices.

Improvements made on farmland may be of three types:

1. Improvements subject to depreciation include farm buildings, silos, fences, tile
   drains, etc., that are made of wood, concrete, brick, masonry, or metal.
2. Improvements that are not depreciable include construction of ditches, soil and water
   conservation expenditures, and cost of clearing land not previously used for farming if
   it was decided to capitalize them instead of deducting them as farm operating expenses. Costs of such improvements are added to the original cost. This total investment is then reduced by the amount of all depreciation previously deducted or
   allowable. If any item has been deducted as an expense, such as soil and water
   conservation expenditures, it cannot be included in the cost basis. Thus it is important
   to have a complete record of all depreciation and capital expenditures during the
   entire period of ownership.
3. Improvements to the farmer's personal dwelling are not depreciable for tax purposes

except for a part used for the farm business office or for hired labor. Their cost should be added to the original investment to determine the basis for gain or loss when sold. (Gains are taxable as capital gains, but losses on one's personal dwelling are not deductible.) If you are age 65 or older when you sell your home, all gain is exempt from income tax if the home sells for not over $20,000. You must have lived in the house five of the last eight years and can have this exemption on only one dwelling.

In all cases where the farmer's personal dwelling is part of the farm that is sold, any gain realized may be postponed for tax purposes if all the proceeds are reinvested in a new dwelling, purchased and occupied by the farmer within one year prior to or after the date of sale of his original dwelling. He has 18 months to build and occupy a new dwelling. If neither is done within the allotted time, any tax on the transaction becomes due in the year of sale.

Installment Sales

In selling a farm, the tax liability on the gain can be spread over a period of two years or more and in many cases can be reduced by the use of the "installment sales method." To qualify for such tax treatment, the payments received in the year of sale must not exceed 30% of the selling price. The remaining income above cost (gain) will be taxable as received in later years. The installment method may be used for any sale of real property and for casual sales of personal property if the selling price is more than $1,000.

Trading a Farm

There is frequently a tax advantage in trading a farm for another farm or other business property. In case of trade all or part of the tax liability is postponed. No gain is recognized for tax purposes unless a difference in cash or nonbusiness property is received in the transaction.

However, in other cases it may be desirable in the long run to sell and pay taxes on the gains in order to get a higher cost basis (including a higher basis for depreciation) on the new farm or business property. This could apply in a situation where an unimproved farm, or a farm on which the depreciation basis has been exhausted or nearly exhausted, is exchanged for a well-improved (with a high depreciation basis) farm or business property. The following example is an illustration. Farmer Brown owns Farm A in which he has a cost basis of $50,000. He has exhausted his depreciation allowances. Farm A has a market value of $100,000. Brown can purchase Farm B for $100,000 or trade Farm A for it. Farm B has a fine set of improvements that can be reasonably valued at one-half the total value of the farm of $50,000 with the other $50,000 allocated to the land. If he trades Farm A for Farm B, he has no tax to pay but retains his cost basis of $50,000 (with no depreciation basis) in the new property.

If he sells Farm A and buys Farm B, Brown will have a $50,000 long-term capital gain ($100,000 sale price — $50,000 cost basis), half of which is taxable. If sold on the installment basis, the gain may be spread over a period of years. If sold with a down payment of $20,000 on the principal and principal payments of $20,000 per year for four additional years, he would have $10,000 long-term capital gain per year, $5,000 of which would be taxable income. Moreover, by purchasing Farm B he has obtained a $50,000 depreciation basis. Since the buildings are not new and obsolescence is a factor, they could quite logically be depreciated on a 10-year straight-line basis, or $5,000 per year.

Thus in the first five years, depreciation would offset the taxable gain and he would be no worse off than if he traded. However, in the next five years, he would have $5,000 per year depreciation that, if he had traded, he would not have had to offset ordinary income.

It is also conceivable to assume that his interest income from the sale of Farm A would be offset or nearly offset by interest expense in purchasing Farm B. Thus, by sale and purchase $25,000 depreciation would be gained to offset ordinary income in addition to the $25,000 depreciation that offsets gain from the sale of Farm A.

TAX PLANNING IN BUYING A FARM

To ensure maximum tax savings at the time of purchasing a farm, the buyer should allocate the total cost of the farm to growing crops, if any; depreciable improvements; dwelling; and land. From a tax management viewpoint the amounts allocated to the various items are handled differently. The cost assigned to the growing crops is an offset (shown on IRS tax form Schedule F) against the selling price of the crop in the year of sale. Of course, the cost basis of the farm is reduced by the amount allocated to the growing crop.

The part of the "cost" allocated to land will not be recovered until the farm is sold since land cannot be depreciated. So, too, the portion allocated to the dwelling is not depreciable if used solely as the buyer's personal residence. A tenant house is depreciable for tax purposes. Cost allocated to depreciable improvements will be recovered through depreciation. Recovery of cost is faster on short-lived improvements than on long-lived ones.

For management and tax purposes the "cost" must be broken down and allocated to each particular structure or improvement. In allocating cost to depreciable improvements, the following procedure may be helpful:

1. Figure the present cost of replacing the improvement.
2. Establish the years of normal useful life.
3. Determine the age of the present improvement.
4. Determine the remaining years of life of the improvement.
5. Compute the present value of the present improvement.

The following is an example of this procedure:

1. Replacement cost of barn—$7,500.
2. Useful life of new barn—25 years.
3. Age of present barn—15 years.
4. Remaining life of present barn—10 years.
5. Value of present barn—10/25 of $7,500 = $3,000. No salvage value need be claimed, as cost of removing old buildings are generally about equal to salvage value.

Another guide in allocating costs is the reasonable insurance values of insurable property. Care should be taken to see that in the final allocation the amount allocated to the bare land represents a reasonable value for similar land in the community.

The proper allocation of cost may help determine the price a buyer will pay for the farm. This is particularly true where the buyer is looking to future farm income after taxes to pay off the purchase price.

Closely related is the manner of payment of the purchase price. In computing taxable income, the buyer deducts interest payments, but not payments on principal. The seller treats interest as ordinary income, while principal payments in excess of his cost basis are capital gains.

MANAGING INCOME FOR MAXIMUM SOCIAL SECURITY BENEFITS

Many farmers may wish to increase their net farm income to the maximum ($14,100 in 1975) in order to secure larger social security benefits (maximum can be increased at any time by Congress). In so doing, income taxes will automatically be increased. However, it may be desirable to do this to gain the additional retirement, disability, and death benefits. These farmers are more interested in methods of increasing rather than decreasing their taxable income.

Some of the methods for increasing income are renting and operating additional land, intensifying and expanding present enterprises, adding new enterprises. electing to report sales of forest products as ordinary income, selling more young stock, doing custom work or other off-farm work, and contracting pasturage and services together, so that the income is recognized as self-employment income and not considered to be rental income.

Where choice of method of handling certain items of expense is optional, farmers may choose the method that gives the smaller deductions. Examples of this would include shifting from a rapid depreciation method for improvements, machinery, etc.; electing to treat soil and water conservation costs as capital investments rather than as current operating expenses; disposing of some depreciable capital items to reduce the total depreciation deductions; and in general, reducing operating costs to a minimum without impairing operating efficiency.

TAX MANAGEMENT TIPS

1. To spread income and reduce taxes, pay reasonable wages to your children for farm work actually done by them as long as there is a true employer-employee relationship. Assign definite jobs or responsibilities, agree on wages, and pay them regularly by check as you would any other employee. Wages paid by parents are not subject to social security tax until the child reaches age 21. Neither are they counted on the child's social security record.

2. A child, under 19 or regularly enrolled in school five months or more in the year or in an accredited on-farm training program, can earn over the personal exemption ($750 in 1975). The parent can also receive an exemption for the child as a dependent as long as he pays over one-half the child's support. This makes possible two personal exemptions, one by the child if he earns over the exemption and files his own return and the second as a dependent of his parent. An individual's total tax-free earnings may be $2,350 as of 1975 (the personal exemption of $750 and low-income allowance of $1,600).

3. An individual's wages are not subject to withholding of federal income tax if he

certifies to his employer that he expects to have no federal income tax liability for the current year and had none for the preceding year.

4. Give income-producing property to children—e.g., land, cattle, machinery—and let them report income from their work and capital. Family partnerships and farm corporations through stock transfers are sometimes used to do this. It is another way to spread family income over the lower brackets. Remember, gifts and partnerships must be legally sound to achieve tax savings.

5. If you are age 63 or 64, postpone income to age 65 to take advantage of the double personal exemption. Persons approaching retirement, however, may want to maintain income as near as possible to the maximum for social security in these years ($14,000 in 1975).

6. At retirement, plan for more income from rents, dividends, interest, and pensions that qualify for the retirement income credit. This may save income tax as earned income is replaced by investment income.

7. Do not hold breeding stock used for production of market livestock too long. By selling sows after only one or a few litters, a higher percentage of hog sales will qualify for capital gains treatment over a period of years and thus reduce taxes.

8. Buy machinery and equipment in years of high income and take the additional 20% first-year depreciation in addition to the regular depreciation allowable.

9. If you are selling or cutting timber, report it as capital gains.

10. Manage sales of farm machinery, equipment, land, and breeding and dairy stock. These can result in capital gains or ordinary losses. Cattle and horses must be held for 24 months or more and other livestock for 12 months or more to qualify for capital gains treatment. Other depreciable items must be held more than six months.

11. Plan personal deductions. Many medical expenses or contributions that are normally spread over two years can be paid in one year and itemized as deductions. In the next year, the standard deduction may be taken if higher than actual deductions. Be sure to choose the largest of (a) the standard eduction, (b) the low-income allowance, and (c) your actual itemized deductions.

12. Plan to have enough income to use up personal deductions and exemptions that are allowed (see Smith example cited earlier).

13. Avoid wide fluctuations in income from one year to the next (see Jones example cited earlier).

14. Understand the effect of rapid or accelerated depreciation that is permissible to use on newly acquired machinery, equipment, and improvements. Decide whether to recover costs quickly or spread them against farm production over a longer period.

15. Installment sales of property can be used to spread income over a period of years and thus avoid high income in one year.

16. Check for loss years in the past. Is there an unused net operating loss deduction? If you file for a refund, be sure to have records to substantiate your claim.

17. **Use the 10% investment credit available on qualified property.**

18. Several alternative courses of action are available to defer payment of taxes. Land development costs can be used as current expenses up to certain limits; additional first-year depreciation and accelerated methods of depreciation are other examples.

19. Filing a tax estimate by January 15 relieves the pressure on farm tax consultants and improves the completeness and accuracy of tax returns, generally saving the tax-payer taxes. In addition, it frees records for use with lenders to establish farm lines of credit early in the year.

20. When gifts of farm products are made to charity, a cash-basis taxpayer can no longer take a fair market value deduction since the raised products have a zero cost basis; therefore, there is nothing to deduct. Depreciable property or appreciable capital gain property is valued at its cost basis to the donor.

TAX REPORTING REMINDERS

1. Be sure that CCC loans are not counted as income twice (in one year when borrowed and next year when crop is sold). Good inventory records will help to prevent this.
2. If you are using the cash method, deduct the cost of purchased livestock lost, strayed, or stolen or that died during the year.
3. If you are using the accrual method, deduct all purchases of livestock. Make a "livestock number check" to see that the total number purchased, born, and in the beginning inventory equals the total number sold, died, butchered, and in the ending inventory.
4. Deduct the farm share of all auto and truck expenses including licenses, insurance, etc.
5. Deduct as much expense of auto, utilities, telephone, etc., as actually used in the farm business (half is not enough in many cases).
6. Take allowable depreciation on improvements; machinery; equipment; and purchased breeding, dairy, and draft animals.
7. Keep records to ensure deduction of easily overlooked items such as farm magazines, farm organization dues, bank service charges, business trips, portion of dwelling used for farm business, household supplies used for hired help, and cash outlay to board hired workers.
8. Itemize on bank deposit slips all gifts, borrowings, savings bonds cashed, etc., so that they will not be considered taxable income.
9. Keep records of all medical, dental, and hospital bills, including premiums for accident and health insurance.
10. Establish a charge account at a hardware store, elevator, or other places where considerable business is done during the year. Pay the account by check upon receipt of monthly statements. This prevents the omission of many small expense items that might otherwise be paid by cash and the receipts lost.
11. Keep exact records of date of purchase, cost, and date of sale on all items purchased for resale.
12. Pay bills by check whenever possible. Record all cash expenditures at once in an account book. Always get receipts for farm expenses paid by cash. Check the monthly bank statement against the farm account book.
13. Do not include in income any indemnity for diseased animals if payment has been or will be used to buy "like or similar" animals within two years for replacements.
14. Withhold and deduct social security tax paid on wages of hired help. When withheld tax plus an equal amount representing your employer tax amounts to $200, they must be deposited currently.
15. Do not report as income the capital gains on the sale of your dwelling if you plan within a year to buy or within 18 months to build and occupy another that will cost as much or more than the selling price of your present dwelling.
16. Keep all "paid" receipts, invoices, canceled checks, etc., for at least five years, including checks for payment of income taxes. Receipts for purchase of items on

Form 10.1. Tax Estimate Worksheet (from Income Tax Management for Farmers, North Central Reg. Publ. 2, 1975)

INCOME TAX ESTIMATE WORKSHEET

		Amount to Date	Estimated Rest of Year	Estimated Year's Total
RECEIPTS:				
Sales of products raised* and miscellaneous receipts:				
Cattle, hogs, sheep and wool, etc.		$____	____	____
Poultry, eggs and dairy products		$____	____	____
All crop sales		$____	____	____
Custom work, prorations and refunds, agriculture program payments		$____	____	____
Total sales and other farm income	(1)	$____	____	____
Sales of purchased market livestock	†$____		____	____
Purchase cost (subtract)	‡$____		____	____
Gross profits on sales of purchased livestock	†(2)	$____	____	____
Gross farm profits (Item 1 + 2)	(3)	$____	____	____

FARM EXPENSES:

Labor hired ....$____	Veterinary, medicine ......$____
Repairs, maintenance .... ____	Gasoline, fuel, oil .... ____
Interest ____	Storage, warehousing .... ____
Rent of farm, pasture .... ____	Taxes ____
Feed purchased ____	Insurance ____
Seed, plants purchased .... ____	Utilities ____
Fertilizers, lime ____	Freight, trucking ____
Machine hire ____	Conservation expenses .... ____
Supplies purchased ____	Other ____
Breeding fees ____	Other ____

		Amount to Date	Estimated Rest of Year	Estimated Year's Total
Total cash farm expenses	(4) $____			
Depreciation on machinery improvements, dairy and breeding stock	(5) $____		____	____
Total deductions (Item 4 + 5)	(6)	$____	____	____
Self employment farm income (Item 3 less item 6)	‡(7)	$____	____	____
Net taxable gain from Schedule D (Sales of dairy and breeding stock, machinery and other capital exchanges)	(8)	$____	____	____
Taxable non-farm income	(9)	$____	____	____
Adjusted gross income (Item 7 + 8 + 9)	(10)	$____	____	____
Less larger of 16% of Item 10 (limit $2,600)** or itemized deductions	§§$____		____	____
$750 x ____ personal exemptions***	$____		____	____
Total non-business deductions and exemptions	(11)	$____	____	____
Taxable income (Item 10 less item 11)	(12)	$____	____	____
Estimated income tax (calculated from applicable tax computation table or rates)	(13)	$____	____	____
Estimated self-employment tax (Item 7 x current rate) ..	(14)	$____	____	____
TOTAL TAX (Item 13 + 14)	(15)	$____	____	____
Less Credits: allowable investment credit and carryover, gas tax, income tax withheld and estimated tax paid	(16)	$____	____	____
Estimated tax due (Item 15 less item 16)	(17)	$____	____	____

---

†Omit for accrual method.
‡For accrual method adjust for change in inventory and new livestock purchases.
§Use itemized deductions if larger.
*For accrual method include sales of all livestock.
**Limit for 1975, see current tax regulation for subsequent years.
***Exemption for 1975, see current tax regulation for subsequent years.

which investment credit was taken should be kept for eight years (or until the property is disposed of) as evidence of purchase. For improvements these vouchers should be kept for as long as the property is held.

17. Remember that if you have income subject to tax, every dollar of cost not deducted will result in unnecessary income taxes.

TAX ESTIMATE WORKSHEET

The tax estimate worksheet shown on Form 10.1 can be used throughout the year in planning farm business and tax management strategies. If not used currently throughout the year, its use in November to plan tax savings in December is strongly advised.

11

USE OF RECORDS AND ACCOUNTS IN FARM PLANNING

THE USE OF ACCOUNTS AND RECORDS as a management tool is the primary purpose for collecting them. Records and accounts provide the best information available for detecting business problems and successes. This chapter illustrates how records can be used to help solve business problems and relates to the analysis phase of decision making discussed in Chapter 1.

In Chapters 7 and 8 the efficiency of the business was measured primarily by comparing the business performance of the past year with other years on the same farm or with other farmers in the same year. This is a useful technique, but other comparisons also may be helpful. One of these is to compare the performance with the business plans (budgets) made prior to the accounting period that has just ended. This phase of the decision-making process is called "evaluation and responsibility." At least two elements should be considered in this comparison of plans with performance. First is the test of the decision-making ability of the manager as he compares his plans with the results. If he meets or exceeds his expectations, he can feel successful about management skills. It should not be interpreted that if performance in any one year or for any particular item does not meet expectations, the manager is a failure. To be successful, the manager must be right only a majority of times or for major decisions. Also, conclusions based upon the performance for any one year may be dangerous because of the variable nature of agricultural production and prices. Second is measurement of the efficiency of the business itself. The comparison of performance with plans (rather than the reverse) may be a useful exercise in identifying business problems. Identifying the strong points of the business may be just as important as identifying the problem areas. These comparisons may take much the same form as the analyses previously discussed.

DATA SOURCES

Farm record data can be much more useful if they are designed and kept with planning in mind. First, more detailed records will be kept. The minimum needed for income tax reporting is not sufficient. Records must be added to determine physical input-output coefficients (feed per pound of gain, seed and fertilizer per acre, labor per head and per acre, fuel and repair costs per hour or acre of use, etc.) and price relationships. Not only must they be kept for transactions off the farm with outside agencies but for intrafarm transfers to be recorded and appropriately charged. Second, production techniques and practices will be designed to give meaningful data for planning. Even though farmers cannot afford costly trials, they can (and many do) carry out limited experimentation. For example, a farmer may plant two different varieties of seed, fertilize at different levels and times, feed different qualities of animals, feed different types of feed to the same quality of animals, sell animals at different weights, harvest crops by different methods, produce at different time periods, etc. Comparison of the results from these types of experiments are helpful to farmers in making decisions. The necessity for complete records of these data is very apparent.

Even though each farmer cannot carry out detailed and accurate experiments for any one activity (let alone a large number), farmers as a group have an abundance of information. As information is shared, they are able to evaluate (within nontechnical limits) crop varieties, fertilizer rates, pest control practices, sources of feeder livestock, feeding practices, new production techniques, etc. Observing what other farmers are doing, specifically rather than generally, can be helpful in adding to the knowledge bank for farm planning purposes. Farmers must be able to attach production requirements to each production level, practice, and variety for their data to be useful for planning.

Major sources of information for farm planning are state colleges of agriculture with their agricultural experiment stations and cooperative extension services. These public agencies are charged with responsibility for conducting research and providing information of benefit to farmers and persons in farm-related businesses. Thus much of the new technology comes from this source. New techniques are developed, and information and products from the private sector are evaluated. Information from these public agencies is useful in two areas of planning. First, farmers can make efficiency comparisons between their records and experiment station results. This is another useful approach to identifying farm problems. Second, the data are useful in supplying information not available to the farmer from other sources (or to test the reliability of information from private sources) for planning purposes, particularly where new techniques are involved.

Other major sources of farm planning data are private organizations. Many of the large supply firms (machinery, feed, fertilizer, seed, chemicals, etc.) have developed their own research sections and have a field staff of specialists to service customers and educate the public about the products they sell. Technically trained salesmen call on farmers, and service divisions design facilities, test soils, analyze records, develop farm plans, etc. Sometimes this source is the only one available for new products. Who is better able to furnish technical information about a dealer's product than the manufacturer?

Even though the farmer cannot generate all the data needed for farm planning, it must all be interpreted through him. Each farm manager is unique in his desires, training, intelligence, age, health, capital position, etc. Thus each new bit of information must be interpreted in light of each farmer. His records and accounts are necessary in making this evaluation.

Much of the information coming to farmers from their own records and from neighbors and private and public sources is in the form of single-point estimates; i.e., each tells the result from one combination of inputs and conditions. This often is enough to indicate whether one practice or input level is more profitable than another; but considering all the possibilities, it may not show which is the most profitable. And what good manager would be satisfied with less? Most products can be produced in several ways. For example, corn can be produced from any one of many hybrid seed corn varieties, with or without commercial fertilizer, and can be harvested with a picker or a combine. Hogs can be produced by feeding straight corn or by supplementing with vegetable or animal protein. Also, the level of production for most products is determined by the level of inputs supplied. For example, the yield of corn can be increased or decreased by increasing or decreasing the amount of fertilizer applied or by changing the plant population. Feeder animals can be marketed at heavier weights by increasing the time and amount of feed. Selecting the best production practice and output level requires a consideration of the costs and benefits associated with each incremental change. The selection of the optimum output level or the combination of inputs to maximize business profit is the subject matter of economics. A consideration of the basic principles of production economics is useful in interpreting business records and accounts and in making future plans.

ECONOMIC PRINCIPLES

Economic principles that may be most useful in planning help to answer the following questions:

1. How much to produce?
2. How to produce?
3. What to produce?
4. When to buy and sell?

Question 1 relates to the quantity of inputs (fertilizer, seed, feed, pesticides, labor, etc.) to apply in producing a product (corn, soybeans, market hogs, feeder cattle, etc.). Another way of looking at the level of production is to consider directly the costs of producing different amounts of product, i.e., crop yields and livestock market weights. There is an economic optimum level of production.

Question 2 considers the combination of inputs to produce a certain amount of product. Many products can be produced in different ways. For example, a market hog can be produced by feeding nearly all corn with very little supplemental protein or by feeding a balanced ration. A 100-bushel corn crop can be produced under minimum tillage or by conventional cultural practices. Even though more than one method is available, there is only one least-cost way.

Question 3 considers the product mix of crops and livestock enterprises on the farm. It considers questions of specialization, participation in government programs, use of idle resources, etc.

Question 4 relates to price expectations for the purchase of input factors and sale of products. This topic will not be discussed directly in this chapter but is part of other economic considerations.

The economic principles discussed here revolve around the *law of diminishing returns,* which states: if successive units of one input are added to given quantities of other inputs required in the production of some product, output of product per additional unit of input will reach a point where the addition to product will decline. It is well recognized that crop plant populations can be too large, fertilizer rates can be too high, and livestock feed efficiency declines as the animals reach heavier weights. These all illustrate the reduction of input efficiency as higher levels of production are attained.

The economic principle governing *how much to produce* is: additional inputs should be added as long as the value of the product added is greater than the cost of the additional unit of factor input. Stated another way, output should be increased as long as the value of an additional unit of product is greater than the cost of the additional units of input required to produce it. This assumes a division of the inputs into variable and fixed packages. Variable inputs are increased as output is increased. They are similar to the operating expenses in the income statement. Fixed inputs do not change with output and are similar to the fixed expenses in the income statement. Whether an input is variable or fixed is determined by time and by experimental design. In the short-run planning period such inputs as rent, insurance, taxes, depreciation, and interest are generally the same (fixed) regardless of what is produced, how it is produced, or at what level it is produced; whereas, such inputs as feed, seed, fertilizer, labor, and fuel can be increased or decreased (variable) in an effort to increase or decrease the level of output. However, if one wishes to measure the output-increasing effect of fertilizer apart from other influences, for example, he may wish to hold even the variable type of inputs fixed while changing only the level of fertilization. Likewise if a cattle feeder wishes to determine the economic optimum weight to feed his cattle, he may measure only the increase of feed required to raise them to additional weight levels apart from considering alternative kinds of feeds and feeding systems. Two examples will illustrate the use of this principle in determining the level of production output.

EXAMPLE 11.1

Suppose a farmer, after studying his own crop records and those of his neighbors, having his soils analyzed at the agricultural experiment station, and consulting with his fertilizer dealer, came up with the following fertilizer-yield relationships for a particular variety of corn (columns 1 and 2 in Table 11.1). Assume that cultural practices, plant population, soils, etc., are the same. Note that the increased fertilizer is 40 pounds per acre at each input level (column 3). Next observe (column 5) that the amount of yield increase per additional 40-pound increment of fertilizer decreases from 12 bushels for the first increment to 0 bushels for the last increment. If corn is worth $1 per bushel and fertilizer costs 8 cents per pound, the added fertilizer cost and returns per acre (columns 4 and 6 respectively) can be tabulated. The added returns received for each additional dollar of fertilizer investment is tabulated by dividing column 6 by 3. As long as this figure is greater than $1.00, it pays the farmer to continue to increase his fertilizer application. This is borne out by looking at column 8 which shows the total income above fertilizer cost for each yield level. It is highest at a fertilization rate of 200 pounds per acre, which corresponds to the $1.25 added returns

TABLE 11.1. Fertilizer requirements for producing corn to different yield levels

(1)	(2)	(3)	(4)	(5)	(6)	(7)	(8)
						Returns per $1 additional investment	Income over fertilizer cost
Fertilizer	Yield	Change in fertilizer		Change in yield			
(lb/A)	(bu/A)	(lb/A)	($/A)	(bu/A)	($/A)		($/A)
0	80						80.00
		40	3.20	12	12.00	$3.75	
40	92						88.80
		40	3.20	10	10.00	3.13	
80	102						95.60
		40	3.20	8	8.00	2.50	
120	110						100.40
		40	3.20	6	6.00	1.88	
160	116						103.20
		40	3.20	4	4.00	1.25	
200	120						104.00
		40	3.20	2	2.00	0.63	
240	122						102.00
		40	3.20	0	0.00	. . .	
280	122						99.60

per $1.00 added investment. Conceptually, income is maximized where the added cost is just equal to the added revenue.

EXAMPLE 11.2

A farmer has the problem of deciding at what weight to market his hogs. From his past records he knows the amount of feed required to produce hogs to various weight levels. From outlook material received from the cooperative extension service he anticipates future prices of feed and hogs. Assuming all other input requirements for producing hogs are held constant, he develops Table 11.2.

Column 2 shows the amount of feed required to increase the weight of the hog from the previous weight level shown in column 1 to the weight level opposite the feed quantity, e.g., increasing the weight from 20 to 40 pounds required 36 pounds of feed. Column 3 summarizes the total feed required to produce a hog to each of the weight levels shown in column 1. When the pigs are small, the higher protein and more refined feeds cost more money per unit than at heavier weights when the feed is mostly made up of corn. The per unit feed prices for the type of feed fed at each weight level are shown in column 4. Columns 5 and 6 summarize the values of the feed fed (columns 2 and 3). Column 5 is tabulated by multiplying column 2 by column 4. Column 6 is tabulated by adding the values in column 5. Column 8 shows the value of a hog at different weight levels and is tabulated by multiplying the total weight (column 1) by the price (column 7). Column 9 shows the change in hog value as heavier weights are reached. As long as the marginal hog value (column 9) is greater than the marginal feed cost (column 5), it pays the farmer to feed to higher weight levels. For this situation the optimum marketing weight is reached at 240 pounds. This is verified by the income figures shown in column 10.

Examples 11.1 and 11.2 illustrate some important points about record keeping. To

TABLE 11.2. Feed costs for producing hogs to heavier weight levels

(1) Total hog weight	(2) Marginal feed required	(3) Total feed required	(4) Marginal feed price	(5) Marginal feed cost	(6) Total feed cost	(7) Hog prices	(8) Total hog value	(9) Marginal hog value	(10) Net income over feed costs
(lb)	(lb)	(lb)	($/cwt)			(cents/lb)			
20		15	8.90	$1.34	$1.34	*	: : :	: : :	: : :
40	36	51	6.00	2.16	3.50	($16/ea)	$16.00	: : :	: : :
60	50	101	4.00	2.00	5.50	*	: : :	: : :	: : :
80	60	161	3.75	2.25	7.75	*	: : :	: : :	: : :
100	68	229	3.50	2.38	10.13	*	: : :	: : :	: : :
120	74	303	3.50	2.59	12.72	*	: : :	: : :	: : :
140	80	383	3.25	2.60	15.32	*	: : :	: : :	: : :
160	84	467	3.25	2.73	18.05	*	: : :	: : :	: : :
180	87	554	3.25	2.83	20.88	21.8	39.24	: : :	$18.36
200	90	644	3.25	2.93	23.81	21.5	43.00	$3.76	19.19
220	93	737	3.25	3.02	26.83	21.3	46.86	3.86	20.03
240	97	834	3.25	3.15	29.98	21.1	50.64	3.78	20.66
260	102	936	3.25	3.32	33.30	20.4	53.04	2.40	19.74

*No prices given for these weight levels.

be useful for specifying optimum production levels, data must be kept in such a manner as to associate production levels with particular input levels. Any one yield level experienced on a farm represents only one level of inputs. Other input levels will give different yield levels. If farmers are to generate their own yield estimates, they must be able to identify the different inputs with each output level. If too many inputs are allowed to change at one time, it may not be possible to identify which influenced the production change. Once the data are kept, they must be organized to associate the amount of output increase caused by each one unit of input increase (or the amount of input or cost increase required to increase output by one unit). Thus the step increases cannot be too large.

The economic principle governing *how products are produced* considers cost reduction and cost additions in changing the method of producing a product. As long as the cost reductions are greater than the cost increases, it pays to make the change. Two examples will be used to illustrate this principle.

EXAMPLE 11.3

A farmer has the problem of which ration to feed his feeder pigs that weigh about 100 pounds. From an agricultural experiment station publication he finds that an experiment has been made to determine the quantities of corn and soybean oil meal, representing different protein levels, that would be required to increase the weight of pigs from 105 to 125 pounds. Next he compares these amounts with his experience, as verified by his records, for the level of protein he has been feeding. After making adjustments, he develops Table 11.3 to represent his situation. Columns 2 and 3 are the basic data. Any combination of corn and soybean oil meal (SBOM) shown (represented by the different protein levels) will increase hog weights from 105 to 125 pounds.

TABLE 11.3. Corn and soybean oil meal requirements for producing pigs from 105 to 125 pounds

(1)	(2)	(3)	(4)	(5)	(6)	(7)	(8)	(9)
Protein	Corn	SBOM	Total feed	Decrease — Corn*		Increase — SBOM†		Total feed cost
(%)	(lb)	(lb)	(lb)	(lb)		(lb)		
10	88.50	2.64	91.14					$1.93
				12.63	$.253	1.52	$.091	
11	75.87	4.16	80.03					1.77
				7.20	.144	1.42	.085	
12	68.67	5.58	74.25					1.71
				4.80	.096	1.40	.084	
13	63.87	6.98	70.85					1.70
				3.43	.069	1.43	.086	
14	60.44	8.41	68.85					1.71
				2.50	.050	1.44	.086	
15	57.94	9.85	67.79					1.75
				1.89	.038	1.51	.091	
16	56.05	11.36	67.41					1.80
				1.48	.030	1.67	.100	
17	54.57	13.03	67.61					1.87

Source: Robert Johnson, Ph.D. thesis, Department of Animal Science, Iowa State Univ., 1965.
*Corn price = 2 cents/lb.
†SBOM price = 6 cents/lb

1. Each combination of corn and SBOM increased the pig weight by 20 pounds.
2. The amount of corn decreased (column 5) as SBOM was increased (column 7) by one-unit drops.
3. The decrease in corn becomes smaller as more SBOM is substituted in the ration for corn (column 5).
4. Feed cost is minimized (column 9) when the cost decrease in corn (column 6) is no longer greater than the cost increase of SBOM (column 8).
5. Feed cost is minimized (column 9 at 13% protein) at a level where total feed fed (column 4) is not at its smallest amount (16% protein). In this example the feed cost is minimized at 13% protein but the results are such that the farmer could use either the 12, 13, or 14% level of protein in the ration without much penalty.
6. The least-cost ration would shift with changes in either the price of corn or the price of SBOM.

EXAMPLE 11.4

A farmer has the problem of how much to invest in laborsaving equipment for feeding his cattle. The number and kind of cattle or type of feed will not be affected. He is interested in three different systems. Considering his present system, this gives four comparisons. For each system he develops fixed and variable costs, which he arrives at after consultation with the industries from which he is buying the equipment and with the county agricultural agent. Table 11.4 summarizes his findings.

System A is the present system. The annual costs for A are variable costs. Systems B, C, and D are new systems. The annual costs for these are their variable costs plus the additional fixed costs greater than A that these systems would have. Column 5 summarizes this table by showing the amount of labor saved per additional one dollar annual cost by changing from one system to the next. If the opportunity cost of labor (the highest return that can be received for labor in an alternative employment) is less than $.80 per hour ($1.00 ÷ $.80 = $1.25), it will not pay to invest in any of these new systems. If the opportunity cost of labor is over $.80 per hour but less than $1.50 per hour ($1.00 ÷ $1.50 = $.67), it will pay to invest in system B but not C. If the opportunity cost of labor is over $1.50 per hour but less than $2.10 per hour ($1.00 ÷ $2.10 = $.48), it will pay to invest in system C but not in system D. At opportunity labor values over $2.10 per hour, system D is the most profitable.

TABLE 11.4. Annual capital costs and labor requirements for four different systems for feeding cattle

System	(1) Annual cost	(2) Labor/ year	(3) Cost increase	(4) Labor decrease	(5) Labor decrease/ dollar cost increase
		(hr)		(hr)	(hr)
A	$ 500	900			
			$200	250	1.25
B	700	650			
			150	100	0.67
C	850	550			
			250	120	0.48
D	1,100	430			

In Examples 11.3 and 11.4, farm records were most useful in furnishing a base of comparison to keep the data relevant and a point by which to measure improvements.

The economic principle governing the *optimum combination of crops and livestock to produce* considers the income added over the income lost as one product substitutes for another in competition for farm resources. There are three major relationships to observe in this regard:

1. *Complementarity.* The income from a product already being produced is increased as a result of adding a new product that likewise has a positive net income. This usually happens where the new product furnishes some needed input to the old product. Examples include nitrogen-fixing legumes, crops that break up disease cycles in existing crops, crops that help control erosion, and livestock that furnish manure to crops. Complementarity may exist only in the long run in some instances. Erosion control in crops is an example. In the short run (one year consideration) the same crops may be highly competitive.
2. *Supplementarity.* The income from a product already being produced is not affected by the addition of another enterprise or activity. This generally is made possible where the added enterprise uses only resources that otherwise would remain idle (go unused). Examples include adding livestock to use winter labor, adding beef cows or sheep to graze a pasture that cannot be used for commercial crops, working off the farm in slack seasons, and selling feed or seed as a side business.
3. *Competition.* The net income from an existing enterprise or activity is reduced as a new enterprise or activity is added or expanded. In this situation there is competition for the same farm resources. An increase of either product decreases the other. It is understandable that all production should take place in this area, which is the competitive range.

In consideration of the above relationships the following rules of thumb are given as a suggested approach to selecting a combination of crops and livestock enterprises:

1. Select the enterprise or activity that promises the greatest income advantage in the planning horizon of the manager and make it as large as resources (including management) will allow. This determination will be made after making an inventory of available resources (acres, soils, machinery, buildings, markets for crops and livestock, credit availability, labor availability, etc.) and studying cost and returns data. Most of the inventory of resources will come from the farm inventory and net worth statement discussed in detail in Chapter 4. This must be the beginning of all planning activity. The planner must know where he is and what resources he has available. The cost and returns data may be from enterprise records kept on the farm (see Chapter 8), experience of neighbors, cooperative extension service personnel or publications, agricultural experiment station publications, or technical material from private agencies. Where past costs and returns data are not available, budgets are necessary for the initial comparisons as well as later projections. Budgeting techniques will be discussed later in this chapter. Making the enterprise or activity as large as resources will allow takes into account the efficiencies of size discussed in Chapter 7.
2. Add complementary and supplementary enterprises until they become competitive with the first enterprise selected. This will require a study of resources available and

those used, disease patterns, soil losses from cropping practices, etc. Records as well as technical information from agronomy, animal science, agricultural engineering, and related disciplines are essential.

3. Add other enterprises or expand complementary and supplementary enterprises to see if their addition or expansion will increase total farm profit. Every addition or expansion of an enterprise at the level of total production will reduce the output of some other enterprise or activity. Thus in order to expand the income of one enterprise, the income from some other enterprise or activity must be reduced. As long as the income added is greater than the income lost, the second enterprise or activity should continue to be expanded. However, in accordance with the law of diminishing returns, as substitution continues, the added net income from the additional enterprises or activities should approach the net income loss from the enterprise being reduced. This comparison requires comparative budgeting or linear programming, the topics to be discussed next. As enterprises and activities are compared, economic levels of output and input combinations should be considered.

4. Other enterprises or activities may be added to reduce farm income variability. However, if there are already three or four major enterprises on the farm, the addition of another will probably not increase income stability. The operator may modify the plan to reduce risks if this is his objective, even though it may mean operating below the economic optimum level over time. His willingness and ability to assume risk may influence the final plan. The degree of risk present varies by enterprise and the management skill of the operator.

The selection of enterprises, the method of production, and the output level all require a system of analysis that will allow the planner to systematically organize and manipulate his data to arrive at a reliable solution to his problem. Problem identification, data sources, and economic logic have all been touched upon. What remains for this discussion is a treatment of analysis techniques and their relationship to records and accounts.

BUDGETING

The budget, a formal plan for carrying out some business activity in the future, must be written down and shows the end results. Budgets contain two parts. One part concerns physical relationships of transforming raw materials (i.e., land, labor, machinery, seeds, beef, wool, cotton, etc.), which requires specifying input-output relationships. The other part concerns price relationships of the inputs and outputs, which allows a specification of costs and returns and calculation of profit. Budgeting as described in this chapter is essentially planning on paper, at which time it is much easier to change the plan than after it has been put into operation. Budgeting may prevent costly mistakes.

All planning is concerned with the future. Often the best guide is the past—if not the past year, then years—which reveals trends and variability as well as levels of use or production. The past combined with outlook information forms the basis for the specification of input-output coefficients and price.

The basic format of a budget is very similar to receipt and expense accounts and the income statement. An understanding of farm records and accounts is essential for planning through the use of budgets.

Budgeting begins with the specification of resource- and operator-inposed limitations. This presumes that the problem has been identified and the objectives specified. Resource limitations might include the land (not just the total acres but the productivity of the soil, precipitation patterns, slope, erosion, weed and insect problems, previous crops, etc.), number and kind of livestock on hand, machinery (number, size, age, and condition of tractors; tillage equipment; harvesting equipment; feed distribution equipment; etc.), buildings (size, type, and condition of grain storage bins; machinery storage; livestock housing; corrals; etc.), improvements (capacity of water wells, tiled land, terraces, fences, etc.), labor (age and condition of operator, availability of family labor, sources of hired labor, etc.), sources of outside capital (credit availability, assets to liability ratio, additional rental properties, etc.), land and livestock tenure arrangements (lease specifications that limit the activities of the operator), available markets, and all other physical restrictions that would set bounds on the planning activity. Operator-imposed limitations might include his education, training, and experience; likes and dislikes; debt aversion; ability to withstand uncertainty of production and income; willingness to work late hours without vacations and weekends; etc.

Budgets may include the total activity of the farm or may be confined to specific parts. The former are commonly called complete farm budgets; the latter are generally referred to as partial budgets. Complete farm budgets are more concerned with the profitability of the total farm than any one part. It is possible that a seemingly very profitable enterprise or activity may be dropped from the farm because it does not fit in with other enterprises or activities. Only in the complete plan can the mix of enterprises and activities be tested and worked out. Partial budgets are used to adjust production methods, change output levels, and test new techniques that do not affect the total farm plan. Once the farm is organized, the partial budget is very useful in developing profitable input-output combinations and output levels for the complete budget. Complete farm budgeting is most often associated with a new farm, addition of a sizable amount of acres to an existing farm, addition or expansion of a livestock enterprise that changes the cropping program, a major shift in supply and demand that affects prices and markets, a change in the tenure arrangements of the farm, a drastic change in government programs, and unexpected changes in the labor supply such as illness and death.

Only variable costs and returns elements are of concern in budgeting. These the planner can control and manipulate. Fixed elements can be carried directly into the summary as they were in previous periods. However, it should be recognized that what may be fixed in one budget may be variable in another. For example, depreciation expense of a tractor on hand would be a fixed cost, whereas depreciation on a tractor the operator is planning to buy would be a variable cost. What is defined as a fixed or variable cost is dependent upon the amount of decision-making power the planner has over that cost. Some costs that normally are fixed may become variable costs over time. Purchases of machinery are a variable cost until after the purchase has been made. Depreciation, taxes, and insurance then become the fixed costs of ownership.

The cash-flow budget has become popular in recent years. This budget anticipates the flow of cash into and out of the business over specified periods of time. Its popularity is associated with changes in the financial structure of the agricultural industry. Due to the magnitude of certain investments, the size of business, small margins of profit, the tightness of credit, etc., farmers have had to tighten the control of their money and make advance arrangements for credit and its repayment. The difference between a cash-flow budget and conventional budgets is illustrated by the tractor example again. Under conventional budgeting, depreciation on a new tractor purchase would be recorded as an expense, whereas in a cash-flow budget the total purchase price would be recorded as an expenditure. Considering that the difference may be $8,000-$10,000, it is easy to see that the net results of the two budgets would not be the same. It may be possible to operate a farm with a negative net farm income in the short run, but it is impossible to operate with a negative cash balance. Also, a particular enterprise or activity may appear to be very profitable in the long run but may have an impossible cash requirement in the short run. Cash-flow budgets do not identify the profitability of a business. They point out the cash-flow requirements only and will indicate the borrowing necessary for the period being considered.

Complete Farm Budget

EXAMPLE 11.5
The budget illustrated here is a basic 240-acre farm located in northwestern Iowa. The length of time since purchase has allowed reduction of the debt load to a comparatively low level. Crops have consisted of corn, soybeans, and meadow on the slopes with oats as a nurse crop. Thirty-five acres will not support row crops. Livestock have consisted of 35-45 litters of pigs and 80-100 head of feeder cattle per year. See Forms 11.1-11.12.

The problem centers around the desire of one of the children to enter the farm business as a partner. Even though the farm has been productive, it will not support another family without expansion. The budget considers the income expansion effects of adding another 120 acres of land, doubling the swine enterprise by adding a swine farrowing house and improving an old barn, increasing cattle feeders by 50 head to about 150 head, and adding 24 head of beef cows to graze the permanent pastures on the two farms.

The nature of this problem requires a look at the total farm before and after the change. Rather than centering on particular year comparisons, it is better to compare the projected future (next five- or ten-year average) with the average of the past several years. Thus there is not a complete connection of the ending inventory and net worth with the beginning inventory and net worth between the two periods. This will cause no problem because no inventory increases or decreases are shown. To conserve space, detailed inventory information is not shown. However, it should be recognized that this information provided the basic framework for the budgeting. Average and projected prices are used rather than those for a particular year.

The budget starts with the *crop plan* (Form 11.1). The crops listed are consistent with previous cropping practices and are selected to provide feed for livestock. Fields lettered A refer to the original farm; those lettered B refer to the 120 acres under consideration for purchase. Not shown are field breakdowns according to productivity or management problems.

Form 11.2 continues the crop plan showing estimated production and disposal of crops. The yields selected are conservative because this area is subject to drought periodically. Note that the total year's crop supply consists of the amount on hand at the beginning of the year plus the added production. At this point it is necessary to determine the livestock requirements before crop sales can be projected. These will be found in Form 11.5. Once tabulated, these amounts are subtracted from the total year's supply. After selecting the normal carryover and subtracting this amount, the quantity available for sale is estimated. It should be recognized that the carryover forms the basis of the ending inventory and net worth statement amount (Form 11.11).

Livestock are interjected in Forms 11.3 and 11.4. Form 11.3 is used to estimate production, purchases, sales, and carryover of livestock. Inventories are held constant as indicated earlier. The major sources of data for this projection are the past records for the farm (birth records, death records, etc.). The anticipated livestock purchases are shown in Form 11.4. Form 11.5 is used to tabulate livestock feed needs. Past feed records of this farm are the best guide to the per head requirements. Once the feed needs have been determined, they are subtracted from the amounts provided by the farm to determine the quantities to be purchased or the quantity available for sale. The farm furnished quantities are obtained from Form 11.2.

Form 11.1. Crop and Livestock Plans

Field number	Acres in field	Year									Budget Year: 1973+	
		a/ 1	2	3	4	5	6				Remarks	
A-1	60	b C	C	S	C	C	S					
A-2	60	C	S	C	C	S	C					
A-3	25	C	C	S	O	M	M					
A-4	25	C	S	O	M	M	C					
A-5	30	S	O	M	M	C	C					
B-1	60	S	C	C	S	C	C					
B-2	20	O	M	M	C	C	S					
B-3	25	M	M	C	C	S	O					
B-4	25	M	C	C	S	O	M					
A-6	35	Continuous pasture									Average annual acres	Percent of tillable land
B-5	23	Continuous pasture										
Acres of corn		c/ 170	170	170	170	170	170				170	
Acres of _SB_		90	85	85	85	85	80				85	
Acres of _Oats_		20	30	25	25	25	25				25	
Acres of hay		50	45	50	55	50	50				50	
Acres of pasture		58	58	58	58	58	58				58	
20% of row crops Acres of set aside		Annual program	decision detail. Corn	based and	upon bean reduced						(50)	

a/ Begin with the first year in the planning period. This usually will be next year.

b/ For each field develop a cropping program consistent with soils, fertility levels, livestock program, conservation practices, etc.

c/ For each year add up the total acres of each crop grown.

d/ Average the acres of each crop over the years in the crop plan.

Form 11.2. Estimated Production and Disposal of Crops

Budget Year: 1973

Name of crop (Column 1)	Amt. on hand (2)	Production Acres (3)	Estimated yield (4)	Total 3 × 4 (5)	Operator share of total production (6)	Total year supply 2 + 6 (7)	To be fed (8)	Feed (9)	Seed (10)	Estimated carry-over Total 9 + 10 (11)	Quant. 7-8-11 (12)	Price (13)	Value 12 × 13 (14)	Time (15)
How obtained		See Form 11.1	5 ÷ 3				See Form 11.5	outlook	outlook	outlook	7-8-11	outlook	12 x 13	outlook
Corn	9,500	140	95	13,300	13,300	22,800	13,300	9500		9500	0			Dec.-March
Soybeans	700	65	33	2,145	2,145	2,845	0		700	700	2,145	2.80	6,006	Dec.-March
Oats	1,000	25	65	1,625	1,625	2,625	1,625	1,000		1,000	0			
Hay	150	50	3,4	170	170	320	170	150		150	0			
Pasture	—	58	2.0	116	116	116	116	0						
Set aside	—	50		50	50	50					50	$.60	3,000	May-Oct
Total		xxxxx	xxxxx	xxxxxx	xxxxxxx	xxxx	xxxxxxxx	xxxx	xxxx	xxxxx	xxxxxx	xxxx	9,006	xxx

Note: If the plan is for next year only, consider the acres of each crop for next year. If the planning period covers several years, the "average annual acres," Form 11.1, may be more appropriate.

Form 11.3. **Estimated Production and Disposal of Livestock**

Budget Year ___1973___

Kind of livestock	No. on hand	Number to be produced	Number to be purchased	Total to account for	Death loss	Home use	To be sold Quant.	Price	Value	Time of sale	Moved out of this class[a]	Carry-over
Cows	20	0	0	20	1		39	60	2340	Jan., Nov		
Bulls	25	25	0	50	1		4	54	216	Feb		↗
Feeders	280	600	0	880	16	2	537	46	24,702	Mar., Sept	45	280
Boars	4	0	4	4	0		4	48	192	Nov		
Steers	104	11	94	209	2	1	102	336	34,272	Aug		104
Heifers	50	6	45	101	1		50	279	13,950	July		50
Calves	24	5	0		1	0	4	225	900		5	19
Total	xx	xx	xx	xxx	xx	xx	xx	xxx	$74,572	xx xxx xx	xxxxx	xx

[a]Example: A heifer on hand at the beginning is counted as a cow in the carryover if she freshened.

Form 11.4. Estimated Production and Disposal of Livestock Products and Cost of Purchased Livestock

Product	No. of producing units	Yield per unit	Total production	Fed on farm	Home use	For sale Quantity	Price	Value
Milk ( )								
Eggs								
Wool								
Total	xxx	xx	xx	xx	xx	xx	xx	$

Cost of livestock to be purchased

Kind	No.	Weight	Price	Amount
Cows	4	300	125	500
Steers	94	450	170	15,980
Heifers	45	425	150	6,750 New
Bull	1	800	400	400
Total			xxx	23,630

Form 11.5. Livestock Feeding Plan

Kind of livestock (Form 11.3)	Period in months and/or days — Pasture	Feeding	No. livestock	Pasture Per head	Total	Corn Per head	Total	Oats Per head	Total	Hay Per head	Total	Concentrates Swine Per head	Total	Cattle Per head	Total
				Ac		bu		bu		ton		cwt		cwt	
Swine 2 litters	3 mo	365	40	.2	8	200	8000	4	160			22.0	880		
Steers fed	60 days	340	105			53	5565	10	1050	.8	84			2.3	241.5
Heifers fed	60 days	310	51			45	2295	8	408	.8	41			1.6	54.1
Cows fed	8 mo	4 mo	25	2	50	2				1.8	45			1.0	25.0
Total feed needs							15,860		1618		170		880		320.6
To be furnished by farm (Form 11.2)							13,300		1675		170		0		0
To be purchased							2,560		0		0		880		321
Price per unit							1.15						6.40		5.80
Value of feed purchased							2,944						5632		1862

Total feed purchased 810,438

224

Form 11.6. Estimated Variable Costs for Labor and Machinery

Budget Year 1973

Operation[a] (Crops — Form 11.2)	No. of acres	Hours per acre (once over)	Number of times over	Total hours	Man labor (show only costs for hired labor) Hours	Price $	Cost $	Powered machinery Units[b]	Price $	Cost $	Implements or equipment Units[b]	Price $	Cost $
Plow	280	.50	1	140				140 hrs	1.40		280 acres	.25	70
Disc & Harrow	280	.28	3	235				235			840	.10	84
Plant: corn & beans	205	.30	1	62				62			205	.30	62
other hay, oats etc.	75	.20	1	15				15			75	.20	15
Cultivate	205	.30	3	184				184			615	.20	123
Fertilize	165	.25	1	42				42			165	.25	41
Harvest: Ear corn	40	.8	1	32				32			40	.75	30
Harvest; Hay	50	2.5	2½	300	200	2.00	400	140			125	.75	94
Overhead 20%								170	1.40				
								1020	1.40				
Total for crops	xxx	xxx	xxxxxxxxxx	xxxxxxxxxx	xxx	xxx		xxx		1428	xxx	xxx	519
Pickup truck & auto			xxxxxxx	xxxxxxxxxx				mi 15,000	.05	750			
Livestock (Form 11.3)	No.	Each						Powered & nonpowered					
Hogs	40								17.50	700			
Steers & heifers	156								3.50	546			
Cows	25								3.00	75			
Overhead													
Total for livestock		xxxx				xxx		xxx		1321	xxx		
Grand total		xxxx				xxx		xxx		3499	xxx	xxx	Total 4018 / 519

[a] The requirements for each crop can be itemized separately or common operations can be grouped over all crops. For example plowing, disking, harrowing, cultivating, etc., can be added for all crops.

[b] Generally, for powered machinery use hours and for implements use acres.

Form 11.7. Estimated Costs for Crop and Livestock Production Inputs and Taxes

Budget Year 1973

Selected crop production expenses — Form 11.2

Crop	Acres	Seed and plants purchased Rate	Amount	Price	Cost	Fertilizer Rate	Amount	Price	Cost	Chemicals Amount	Price	Cost	Custom work Rate	Cost	Other Rate	Price	Cost
Corn	140	15#	38 bu	$20.00	$760	120-80 -50		$1760	$2464	2.00	$2.00	$280	$8.00	$800	spraying	$5.05	$665
Soybeans	65	1.2	78 bu	4.00	312	none			0		2.00	130	5.00	325	shelling		
Oats	25	30	75 bu	1.25	94	50-60 -0	9.00		225				4.00	100			
Hay	25	75.3	175/75	.50	188	lime	3.50		98		2.00	50			twine		150
Seeding	50		50 au	5.00	250	0			0								
Pasture	58					50-20 -10	5.00		290	1.00		58					
Total	xxx	xxx	xxx	xxxx	1604	xxx	xxxx	xxxxx	3067	xxxxx	xxxx	518	xxx	1225	xxx		815

Livestock expense — Form 11.3

Item	Cost
Veterinary	$134/
Medicines	
Registration	
Shearing	
DHIA	
Breeding	
Other	
Trucking	768
Total	2109

Increase or decrease in real estate and personal tax estimate from previous year

Item	Inventory value change	Assessed value	Item	Inventory value change	Assessed value
Land	$72,000	$18,440	Hogs	$	$
Buildings	5,000	1,350	Grains		
Machinery					
Equipment					
Dairy cows					
Beef cows			Feeds		
Heifers					
Bulls			Seed		
Feeder cattle			Total	xxx	20,790
			Tax (total x levy)	.08	$ 1,663

226

Form 11.8. Labor Requirements and Distribution for Crops, Livestock, and Farm Overhead

Budget Year _1973_

	Number of units	Labor needs Per unit	Labor needs Total	Jan.	Feb.	Mar.	Apr.	May	June	July	Aug.	Sept.	Oct.	Nov.	Dec.
Crops (Form 11.2)															
Corn	140	4.5	630	6.3	6.3	12.6	94.5	94.5	88.2	37.8	25.2	25.2	126.0	100.8	12.6
Soybeans	65	4.0	260	2.6	2.6	5.2	31.2	44.2	44.2	46.8	72.8	39.0	18.2	2.6	2.6
Oats	25	2.0	50	0.5	0.5	7.0	10.0	3.0	1.5	20.0	4.0	2.0	0.5	0.5	0.5
Hay	25	2/ton	340	0.0	0.0	10.2	13.6	13.6	119.0	119.0	40.8	17.0	3.4	3.4	0.0
Rotation	50	1.0	50	0.0	0.0	15.0	10.0	2.5	2.5	5.0	5.0	5.0	5.0	0.0	0.0
Pasture	56	1.0	56	0.0	0.0	11.2	5.6	5.6	5.6	16.8	5.6	2.8	2.8	0.0	0.0
Total direct hours for crops			1386	8.4	9.4	61.2	164.9	163.4	261.0	245.4	101.4	91.0	165.9	107.3	155.7
Livestock (Form 11.3)															
Hogs (2 litters)	40	28	1120	145.6	123.2	112	78.4	67.2	89.6	100.8	89.6	89.6	56.0	67.2	100.8
Feeder cattle															
Steers	105	8.0	840	84.0	84.0	84.0	67.2	58.8	84.0	84.0	84.0	42.0	42.0	42.0	84.0
Heifers	51	7.0	357	35.7	35.7	35.7	28.6	25.0	35.7	35.7	35.7	17.8	17.9	17.8	35.7
Cows	25	10.0	250	30.0	30.0	30.0	35.0	20.0	15.0	12.5	10.0	10.0	10.0	17.5	30.0
Total direct hours for livestock			2567	295.3	272.9	261.7	209.2	171.0	224.3	233.0	219.3	159.4	125.9	144.5	250.5
Overhead labor for crops & livestock		25/ac	1000	100	50	50	100	100	50	100	200	100	50	50	50
Total labor required for farm			4953	404.7	532.3	372.9	474.1	434.9	535.3	578.4	520.7	350.4	331.8	301.8	316.2

Form 11.9. Labor Availability and Requirements Chart

Budget Year _1975_

Hours per month	Jan.	Feb.	Mar.	Apr.	May	June	July	Aug.	Sept.	Oct.	Nov.	Dec.	
600													Labor Available
550													Labor Required
500													Hired Labor
450													
400													
350													
300													
250													
200													
150													
100													
50													

Total hours hired ≈ 300 @ 2.00 = 600 mostly for hay harvest.

228

Form 11.10. Added Investments and Annual Cost Changes

Budget Year 1973+

Item	Month required	Cost	Trade-in	Added cost	Depreciation Before	Depreciation After	Depreciation Difference
Machinery:							
Tractor 100 hp	Mar.	$9,000	$2,000	$ 7,000	$400	$900	$ 500
Plow, 5 – 16"	Mar.	1,400	400	1,000	100	140	40
Planter, 6-row	Mar.	2,500	300	2,200	150	250	100
Cultivator, 6-row	May	1,300	500	800	90	130	40
Baler	June	2,500	---	2,500	0	250	250
Feed mill	Jan.	2,200	---	2,200	0	220	220
Feed wagon	Jan.	1,000	---	1,000	0	100	100
Misc. livestock		1,000	---	1,000	0	200	200
				17,700			1,450
Building improvements:							
Farrowing house	Sept.	5,800	---	5,800	0	400	400
Barn modifications	Aug.	2,000	---	2,000	0	135	135
Corral fences	Aug.	1,500	---	1,500	0	150	150
Pasture fences	June	800	---	800	0	80	80
Miscellaneous		1,000	---	1,000	0	200	200
				11,100			965
				$28,800			$2,415
Livestock:							
Beef cows (24)	Jan.	6,000		6,000			
Beef feeders (50)	Oct.	8,000		8,000			

Machinery costs are tabulated in Form 11.6. Farm records are again the best source of machine requirements, but farmers as a whole are less likely to have this kind of information. However, an acreage expansion of past machinery costs would provide a useful guide and base of comparison. Several approaches are available to estimate machinery cost. For crops a functional approach that crosses enterprise boundaries is used. An enterprise approach is used for livestock. In both cases only variable costs are shown. Fixed costs of depreciation, interest, housing, insurance, and taxes are shown separately.

Crop and livestock production and machinery hiring costs are shown in Form 11.7. Typical levels of seed, fertilizer, chemicals, veterinary costs, medicines, etc., are used. This farm does not have a combine, so custom rates are applied. Also included in this table are increased tax estimates. Once again past practices as recorded in the farmer's records provide most of the basis for estimates.

Labor requirements are estimated in Forms 11.8 and 11.9. Form 11.8 is used to estimate individual enterprise requirements by month. Overhead requirements are more flexible than these projections would indicate. In Form 11.9 the farm requirements are compared with the labor supply. Small differences do not necessarily mean that operator labor will be in excess or that additional labor must be hired. Large differences indicate potential problems. In this instance labor appears to be adequate except for hay harvest and perhaps plowing. Also, it would appear that the son could do some off-farm work for wages during the fall.

Additional investments are summarized in Form 11.10. The larger acreage will require having larger machinery, and the expanded livestock production will require some new equipment and facilities. Since there will be some trade-ins, the added cost and the depreciation difference columns are important for the summary forms.

The *net worth statement,* Form 11.11, summarizes the "before and after" investments and liabilities. Sources of different entries are shown to the right of each; most of these are found in the budgets just discussed. The net capital ratio is still over 2:1 after the change and should be considered safe.

The projected *income statement,* Form 11.12, summarizes the total budgeting effort. Again the sources of elements are shown on the right. In this case net income has been increased by nearly $5,000 after payment of interest, but two families must live from this. Also, if total principal payments are greater than depreciation, the difference must be paid from net income.

An *income distribution analysis* is shown in Table 11.5. Here the net income is distributed to labor and capital. Using an annual wage of $6,000, the return to capital is tabulated as a residual. The overall return to capital has been increased by 0.2%. This means that the return on the added investment is 0.5% greater than on the original investment. From this standpoint it looks favorable, but the total interest return still may be too little (6.6% return compared to a 7.0%) to justify the investment. The $6,000 labor return may be too low to attract the son. The actual decision must be based upon the goals of the family, the anticipated capital gains of the investment, the willingness to sacrifice, etc. This analysis indicates that it is an economic feasibility but not a glamorous investment opportunity.

Form 11.11. Net Worth Statement

Asset[a]	Value Beginning	Ending Plan 1	Ending Information source
Liquid			
Cash on hand – checking	3,400	3,400	Bank statement
Life ins., cash value	2,500	2,500	
Bonds, present value			
Notes and accounts receivable			
Savings	5,000	0	Savings book
Feeds	9,800	14,100	Inventory form 11.2
Seed and supplies	1,960	1,960	Inventory
Feeder livestock			
Cattle	19,600	28,010	Inventory form 11.3
Hogs	5,200	10,100	Inventory form 11.3
Sheep			
Poultry			
Working			
Machinery and equipment	13,500	31,200	Depre. schedule
Movable buildings	400	400	"
Breeding livestock			
Cattle	0	6,400	Inventory form 11.3
Sheep			
Fixed Farm B 160 acres			
Land 240 acres @ 440	105,600	70,400	Account book
Farm B		105,600	
Buildings attached to land	2,400	15,100	Depre. schedule
Total assets	144,360	289,170	

Liability	Value Beginning	Ending Plan 1	Ending Information source
Short-term			
Notes payable			
Cattle	15,000	21,000	Credit account
Accounts payable			
Feed	1,000	2,000	Credit account
Cash rent due			
Intermediate-term			
Machinery chattels	0	16,000	Credit account
Equipment		1,000	"
Livestock: chattels loans		5,000	"
Increasing loan		5,000	"
Long-term			
Real estate mortgages			
Farm A	20,000	20,000	Credit account
Farm B		72,000	
Building modifications		2,000	Form 11.10
Total liabilities	36,000	144,000	
Net worth	128,360	145,170	

[a] The items named here are those normally included with the farm business. There is some arbitrariness about the classification of assets and liabilities used. It may be desirable to inventory also personal (family) assets and liabilities and other nonfarm businesses in order to have a complete picture of the assets and liabilities of the operator and his family.

[b] These three columns can be used alternatively for operator, landlord, and total farm rather than for different plans being tested. The purpose for including the landlord is to obtain estimates for comparative purposes.

Form 11.12. Net Farm Income Statement

DEBITS   Budget Year *1973*

Item[a]	Past year[b]	Budget year Plan 1	Budget year Information source
1. Cash operating expenses			
2. Labor	671	600	*Forms 11.8, 11.9*
3. Truck and machine hire	814	1125	*Form 11.7*
4. Fuel and oil (farm)	1133	⎫	
5. Auto expense (farm)	420	⎬ 4018	
6. Machine repair and maintenance	1781	⎭	*Form 11.6*
7. Building repair and maintenance	489	588	*Record projection*
8. Crop expense (seeds, pesticides, etc.)	1752	2937	*Form 11.7*
9. Crop expense (fertilizer)	2275	3067	*Form 11.7*
10. Livestock expense (nonfeed)	1507	2109	*Form 11.7*
11. Utilities (farm)	449	550	*Record projection*
12. Miscellaneous	220	250	" "
13.			
14. Total (1 thru 13)	11,311	15,244	
15. Livestock feeds: *Poultry*	1640		*Form 11.5*
16. Cattle	1500	1862	
17. Hogs	2900	5632	
18. *Corn*		2944	
19. Total (15 thru 18)	6040	10,438	
20. Livestock purchases			*Form 11.4*
21. Cattle *feeders*	15,500	22,730	
22. Hogs *boar*	250	500	
23. *Bull*		400	
24. *Poultry*	150		
25. Total (20 thru 24)	15,900	23,630	
26. Cash fixed expenses			
27. Taxes	1343	2923	*Form 11.7*
28. Interest	2715	10,920	*Credit account*
29. Insurance	552	680	*Record projection*
30. Rent			
31.			
32. Total (26 thru 31)	4610	14,523	
33. Total cash expenses (14+19+25+32)			
34. Noncash expenses			
35. Depreciation, buildings	1077	2042	*Deprec. schedule*
36. Machinery and equipment	1569	3019	" "
37.			
38. Total (35 thru 37)	2646	5061	
39. Decreased inventory, livestock			
40. Liquid assets			
41.			
42. Total (39 thru 41)			
43. Total business debits (33+38+42)	40,507	68,896	
44. Capital item purchases			
45. Machinery and equipment			
46. New improvements			
47.			
48. Total (45 thru 47)			
49. Total cash expenditures (33+48)			

[a] Form adapted after the one used by the Iowa Farm Business Association.

[b] If complete farm records are not available, show figures used to calculate the latest income tax.

[c] These three columns can be used for the operator, landlord, and total farm for one farm plan rather than for different plans being tested.

Form 11.12. (continued)

Budget Year *1973+*

Item	Past year[b]	Value Budget year Plan 1	Information source
50. Livestock sales			
51. Cattle	31,700	49,122	*Form 11.3*
52. Hogs	13,700	27,450	" "
53.			
54.			
55. Total (50 thru 54)	45,400	76,572	
56. Livestock product sales			
57. Milk			
58. Eggs	2300		
59. Wool			
60.			
61. Total (56 thru 60)	2300	—	
62. Crop sales			*Form 11.2*
63. Corn			
64. Soybeans	3600	6006	
65.			
66. Diverted acreage payments	2380	3000	
67.			
68. Total (62 thru 67)	5980	9006	
69. Other			
70. Machine work *and labor*	400	1400	*Record projection*
71. Refunds	562		*Record projection*
72. Dividends		800	" "
73.			
74.			
75. Total (69 thru 74)	962	2200	
76. Profit or loss, sale of capital assets			
77. Cash income (55+61+68+75+76)	54,642	87,778	
78. Noncash receipts			
79. Increased inventory			
80. Liquid assets			
81. Livestock			
82.			
83. Total (78 thru 82)			
84. Farm products used in home	210	380	*Form 11.3*
85. Total business credits (77+83+84)	54,852	88,158	
86. Capital asset sales			
87. Undepreciated value, machinery			
88.			
89. Total (86 thru 88)			
90. Total cash farm income (77+89)			
91. Gross profits (77+83-42) + (84-19-25)			
92. Net cash farm income (90-49)			
93. Net farm income (85-43)	14,345	19,262	

233

TABLE 11.5. Distribution of income and return on added
land investment

Net farm income	$ 14,345	$ 19,262
Plus interest paid	2,715	10,920
Minus family labor	1,100	. . .
Adjusted net farm income	$ 16,960	$ 30,182
Minus operator labor:		
Father	$ 6,000	$ 6,000
Son		6,000
Return to capital	$ 9,960	$ 18,182
÷ Capital managed	$164,360	$289,170
Percent return to capital	6.1	6.3
Return on added investment:		
Added return to capital		$ 8,222
Added capital managed		$124,810
Percent return		6.6

Partial Budget

EXAMPLE 11.6

Joe Farmer owns a 160-acre farm. In addition to feeding 100 head of cattle, he still finds time to work off the farm part time. He has been realizing an average of $8,000 net farm income plus another $3,000 from nonfarm sources. He is considering additional investments that will employ him more fully on the farm. Near his present farm, an 80-acre piece of land is for sale which he thinks he can buy for $600 per acre. Also, he has been considering adding 100 head of feeder cattle. Joe does not foresee any investment problems since he has over a 60% equity in his current farm business. However, he does not think he can act on both ideas at the same time. This requires some alternative budget work to estimate the probable outcomes. He must evaluate the alternative returns, repayment capacity, and risk factors.

Partial budgets take the form of:

A. Added incomes
  1. Increased sales from new activity
  2. Decreased costs from replaced activity
B. Added costs
  1. Increased costs from new activity
  2. Decreased sales from replaced activity
C. Net income increase = A − B

This approach seems appropriate for this problem.

Alternative I—80-Acre Land Purchase

This alternative is the purchase of 80 acres of land for $600 per acre, which can be financed with a 25-year amortized loan at 7% interest, and the purchase of some larger equipment to raise 80 acres more corn (or soybeans, assuming the same net profit). Land taxes are $7 per acre. Average equipment life is 10 years and the interest rate is 7% (3.5% on the original investment). Labor requirements for corn are now 6 hours per acre, but with the larger equipment this can be reduced to 5. The present return above variable costs per acre is $70, but with the reduction in machine time this can be increased by $2. The average annual capital can be secured at 7% interest. The current acres in row crops (principally corn) are 140. Income from off-farm work is $2 per hour. The land and equipment investment summary and tabulations for Alternative I are shown in Table 11.6.

Alternative II—100 Head of Feeder Cattle

The present beef feeding enterprise involves the use of an upright silo with cattle being fed by hand from a wagon. The proposed system involves adding an upright silo with an unloader, adding an unloader to the present silo, and purchasing a self-unloading wagon. The feed yard is large enough and the water supply is sufficient, but additional feed bunks must be built. Currently 8 hours of labor are required per feeder, but with added automation this could be cut to 6 hours. Past returns per $100 feed fed have averaged $145, but the future looks a little brighter and $150 seems probable. Joe Farmer has been working off the farm for $2 per hour. Money can be borrowed at 7% interest. The livestock and feed capital requirement is $200 per head. The investment and input summary and tabulations for Alternative II are shown in Table 11.7.

Decision Making—Alternative I, Alternative II, or Present

Both alternatives promise a higher net income than the present off-farm job but neither may be as reliable. Also, the increase is not great in either

TABLE 11.6. Summary for Alternative I

Item	Present system	Proposed system	Change
Land and buildings	$96,000	$144,000	$48,000
Machinery:			
Tractor	5,000	7,000	2,000
Plow	800	1,100	300
Disk	600	900	300
Planter	400	800	400
Cultivator	500	700	200
Harvest equipment	1,100	1,900	800
Total	$8,400	$ 12,400	$ 4,000
Corn acres	140	220	80

Tabulations — Alternative I	
A. Added incomes	
1. Corn sales (80 A x $72/A)	$5,760
2. Machine cost reduction (140 A x $2/A)	280
	$6,040
B. Added costs	
1. Land interest ($48,000 x .07)	$3,360
Land taxes (80 A x $7)	560
Machinery interest ($4,000 x .035)	140
Machinery depreciation ($4,000 ÷ 10)	400
Production capital (80 A x $16 x .07)	90
2. Off-farm labor reduction	
[(80 A x 5 hr/A) − (140 x 1 hr/A)] $2	720
	$5,270
C. Net farm income increase (A − B)	$ 770
Net cash farm income after mortgage payment	$ 11
(Total payment $4,119 − $3,360 interest = $759)	

TABLE 11.7. Investment and input summary for Alternative II

Item	Present system	Proposed system	Change
Capacity (head)	100	200	100
Returns/$100 feed fed	$ 145	$ 150	$ 5
Investments:			
Housing, fencing, concrete	$ 4,000	$ 7,500	$ 3,500
Feed storage and unloaders	6,000	18.000	12,000
Feed wagon	200	1,500	1,300
	$10,200	$27,000	$16,800
Utilities	$ 130	$ 250	$ 120
Labor (hr)	800	1,200	400

Tabulations — Alternative II	
A. Added income	
1. Beef returns over feed costs (100 hd x $50/hd)	$5,000
B. Added cost	
1. Depreciation ($16,800 ÷ 15-yr life)	1,120
Interest on investment ($16,800 x .035)	588
Interest on cattle (100 hd x $200/hd x .07)	1,400
Taxes on average investment [(16,800/2) x .27 x .08]	181
Repair ($16,800 x $.02/$1 invested)	336
Utilities	120
2. Reduced off-farm labor income (400 hr x $2)	800
	$4,545
C. Net income increase (A − B)	$ 455

case. On the other hand, there may not be the personal satisfaction or opportunity to grow with the off-farm work. The land investment opportunity (Alternative I) has the highest net income increase and probably is more certain and stable than the cattle investment (Alternative II). But after mortgage payments are made, the amount of cash income increase is near zero. The land does offer the opportunity for later livestock expansion, and another 80 acres that is so conveniently located may not be for sale soon. Also, land may have higher long-term capital gains than feeder cattle. With the feeder cattle expansion other planning horizons and price projections may give more or less favorable results. Another $5 per head would add $500 to net income. If a 20-year investment life had been used on facilities, the depreciation expense would have been reduced by $280. Some of these matters must be evaluated in terms of the farmer's ability and willingness to cope with uncertainty, his values and goals, and elements not entirely economic in nature. But the economic elements are good to know.

Cash-Flow Budget

The cash-flow budget traces the flow of cash into and out of the business over a period of time, usually in blocks of one year each. This type of budget differs from other budgets primarily in two regards. It deals only with cash transfers and breaks the year down into periods usually of one month each. This allows the farm planner to anticipate his monthly cash productive balance and his credit needs and repayment schedule. Due to business size and credit needs of the modern farmer this becomes extremely important.

EXAMPLE 11.7
The complete farm budget used in this chapter (Forms 11.1-11.12) will form the basis of the cash-flow budget illustration. Some adjustments must be made since an average situation was budgeted rather than the expectations for the next year. To conserve space, bimonthly flows, which may be sufficient for some farmers, will be used rather than monthly. The cash-flow budget is shown in Form 11.13. Note the following:

1. The organization is similar to the income statement.
2. The cash amounts for the past year(s) form the basic background for the coming year's cash flow.
3. Deviations from the past are obtained from complete or partial farm budget projections. Compare the total projected amounts with the income statement amounts shown in Form 11.12.
4. Production type expenditures are included as well as capital purchases and principal and interest payments on loans committed in previous years.
5. If family living and investment expenditures come from the farm, these transfers also must be included.
6. The monthly allocations are obtained from records of past years or budget materials. Projected sale and purchase dates are shown in Forms 11.2-11.10.

The credit summary is shown at the bottom of Form 11.13. Once the monthly cash flows are obtained, the surplus and deficit months can be identified. Since it is inadvisable to have a negative bank balance, loans must be obtained in sufficient amounts to keep the bank balance positive. It is good business practice to keep a minimum amount in the bank at all times as a cushion against the unpredictable. In months when there is a surplus, loans can be repaid and interest payments made. Note that the beginning and ending cash balance agrees with the amounts on the net worth statement (Form 11.11). Also, the outstanding borrowings approximately agree ($34,975 compared to $34,000) with the liability difference between the two years if the land purchase loan is excluded in the net worth statement as it is in the cash-flow budget. This presumes that the land loan was obtained in a previous year or was handled completely separate from the cash-flow budget. These two approximations were arrived at separately but were from the same data.

LINEAR PROGRAMMING

Linear programming is a means for solving complex farm problems with the aid of an electronic computer. Even though this has many applications for the farmer, its primary use has been to help determine the economic optimum combination of products to produce. Acres of crops and numbers of livestock are specified as well as variable input purchases (labor, feed, fertilizer, feeder livestock, etc.), intrafarm transfers (grain to livestock, calf raising to calf feeding, meadow to hay or grazing, etc.), and the marginal values of the restricted resources (what an additional acre of land, head of livestock, hour of labor, etc., would return in net income) are shown. Linear

Form 11.13. Cash-Flow Budget

	Last year	Total projected	Jan.-Feb.	March-April	May-June	July-August	Sept.-Oct.	Nov.-Dec.
Income (from all sources)	$	$	$	$	$	$	$	$
Livestock sales, cattle	31,700	49,122				48,222	900	
hogs	13,700	27,450	216	12,351	1,170		12,351	1,362
Livestock product sales	2,300							
Crop sales: soybeans	3,600	6,006		1,960			4,046	
Government payments	2,380	3,000			1,500		1,500	
Machine work income	400	1,400	200	200			500	500
Other income	562	800	400			400		
Total inflow	54,642	87,778	816	14,511	2,670	48,622	19,297	1,862
Expenditures								
Feed, commercial	6,040	7,494	1,249	1,249	1,249	1,249	1,249	1,249
Feed, grain		2,944			2,944			
Livestock purchases--cattle feeders	15,500	22,730						22,730
cows, bulls, boars	250	6,900	6,000	400			500	
New purchases--machinery improvements	5,000	28,800	3,200	10,700	4,100	3,500	5,800	1,500
Labor	671	600			300	300		
Machine & equipment repair	1,781	4,018						
Gas, fuel, oil	1,133		201	803	1,205	603	804	402
Machine hire	814	1,225				100	325	800
Auto operating	420	above						
Utilities	449	550	100	100	83	83	84	100
Fertilizer and lime	2,275	3,067			3,067			
Other crop expense	1,752	2,937	300	1,604	518	150		365
Livestock expense	1,307	2,109	252	301	252	591	462	251
Building repairs	489	588	58	59	59	294	59	59
Taxes, real estate and personal property	1,343	2,923		1,461			1,462	
Insurance	552	680	240		440			
Rent								
Other farm expense	220	250	42	42	41	42	42	41
Living expenses	6,710	12,000	2,000	2,000	2,000	2,000	2,000	2,000
Income tax and social security--state	354	394		394				
federal	1,869	1,968		1,968				
Debt payment due	17,000	17,000	1,000	1,000				15,000
Interest due	2,700	2,715		1,665				1,050
Total outflow	68,629	121,892	14,642	23,746	16,258	8,912	12,787	45,547
Net cash flow (+ or -)	-13,987	-34,114	-13,826	-9,235	-13,588	39,710	6,510	-43,685

Summary of Credit Needs for 1973

	Total projected	Jan.-Feb.	March-April	May-June	July-August	Sept.-Oct.	Nov.-Dec.
Net cash flow (+ or -)		-13,826	-9,235	-13,588	39,710	6,510	-43,685
plus cash available, beg. period	xxx	-10,426	-5,661	-10,149	43,161	12,110	-31,575
less payments on new borrowings this year--principal	xxx				36,700		
interest	xxx				861		
Total cash available (+ or -)	xxx	-10,426	-5,661	-10,149	5,600	12,110	-31,575
Borrowing for $____ balance $____	xxx	14,000	9,100	13,600			34,975
Cash balance, end of period	xxx	3,574	3,439	3,451	5,600	12,110	3,400

programming can be seen as a means of comparing a large number of farm budget alternatives and selecting the best combination.

The word linear arises from the fact that the program does not automatically adjust for changing input-output ratios as the size of the enterprise or activity increases. The particular input-output ratio is held constant until the programmer builds into the model a stop beyond which alternative activities (budgets) are considered.

Thus the problem of diminishing returns is handled by building into the program a different activity for each input-output combination to be tested. For example, corn fertilized at 100 pounds per acre is a different activity from corn fertilized at 200 pounds per acre, and cattle marketed at 1,150 pounds is a different activity from cattle marketed at 1,050 pounds. Likewise, it is possible to test the use of different combinations of inputs to produce a product. Minimum tillage corn is a different activity from a conventional method of production. Sows farrowed in the winter is a different activity from sows farrowed in the spring. And cattle fed silage is a different activity from cattle fed hay. Linearity, then, is not a difficult problem if the programmer understands the mechanics of the programming model and the physical and economic relationships of agricultural production.

The information needed for the linear programming method of farm planning is exactly the same as required for budgeting. However, more attention is given to the specification of restrictions caused by resource limitations and other forces. Unit budgets (the requirements for producing one unit of product) are developed for each activity to be tested; whereas in regular budgeting, activities and input requirements may be combined.

Thus the first step in developing a linear program is to take an inventory of resources. This should be very specific as to the quality of land (whether it will support continuous row crops or must be planted to soil conserving crops, weed problems, flooding or drainage problems, drought problems, etc.); the availability of operator, family, and hired labor by the month, or at least production season; government program restrictions; the types and conditions of buildings and facilities (the number of sows that can be farrowed at a time and during which seasons, number of cattle that can be fed in the bunks, livestock water limitations, etc.); and the kind and condition of machinery; etc. In addition, management restraints should be listed. These may include an upper limit on the number of sows the operator is willing to farrow, whether he is willing to hire additional labor, whether he can obtain or is willing to use credit, if he wishes to take a vacation at a particular time of the year, etc.

The second step is the selection of enterprises and other activities to be considered. The enterprise activities are commonly referred to as unit budgets. They specify the physical resource requirements, particularly for those that may be limiting, and the costs for producing one unit of product. The third step is the selection of prices, which are a part of the budget development process, for the inputs and products. These are then entered in the computer in a prescribed format. The computer expands the unit budgets within the framework of restrictions imposed by management and physical limitations and selects the combination of products that promises the highest net income over variable costs. As with other decisions, only the variable costs are useful in selecting between alternatives. The fixed costs can be subtracted from the value of the program to tabulate net farm income.

As with other farm planning, the importance of farm records is apparent. They are particularly useful in furnishing information for specifying resource restrictions and

alternative points for consideration are *A, B, C, D,* and *E.* The alternative production levels are shown in Table 11.10. The substitution ratio measures the bushels of corn that must be given up to produce one more bushel of soybeans. Since the net profit of soybeans is $1.33 per bushel and corn $.55 per bushel, it is profitable to keep substituting soybeans for corn until the substitution ratio reaches 2.42 ($1.33 ÷ $.55). Since the substitution ratio is less than this when going from Alternative B to C and it is greater when going from C to D, Alternative C is selected to give the most profitable combination (this converts to 27 acres of corn and 73 acres of soybeans). However, in the long-run situation where all prices become variable (new machinery must be purchased, storage facilities built, etc.), the net prices of corn and soybeans drop to $.05 per bushel and the price ratio is 1.00. In this case Alternative B is selected. Going from Alternative A to B is a supplementary situation.

The computer reaches the above conclusions by solving a series of mathematical equations. The above example could be formulated as follows:

$$1.0C + 1.0S + 1.0X_1 = 100 \text{ acres of land}$$
$$0.3C + 0.9S + 1.0X_2 = 80 \text{ hours of April labor}$$
$$0.8C + 0.5S + 1.0X_3 = 80 \text{ hours of May labor}$$
$$0.7C + 0.4S + 1.0X_4 = 80 \text{ hours of June labor}$$
$$1.1C + 0.4S + 1.0X_5 = 80 \text{ hours of October labor}$$
$$1.5C + 0.0S + 1.0X_6 = 80 \text{ hours of November labor}$$
$$54.81C + 35.29S + 1.0X_7 = 4,000 \text{ dollars of operating capital}$$
$$52.25C + 43.89S = Z$$

The symbol *C* refers to acres of corn and *S* to acres of soybeans. X_1 through X_7 are nonuse activities that are added to allow resources to remain idle if that is more profitable. On the right are shown the resource limitations. The corresponding *C* and *S* coefficients show the quantity of that resource required to produce one unit (acre) of corn or soybeans. Representing net profit over variable costs, *Z* is to be maximized.

EXAMPLE 11.10

This example presents a linear programming solution to the complete farm budget illustration in Forms 11.1-11.12. The linear programming model for this farm situation is presented in Table 11.11. To conserve space, not all activities and restraints are shown. A complete list of the activities is given in Table 11.12. A reading of this list will reveal that the activities closely follow the decision-making opportunities on the farm. For example, the best land (Land I) can be planted to a variety of crops (P01-P04); whereas Land II is more restricted to P05. Cattle and swine can be produced in their present facilities (P08, P10, P11) or in new facilities (P09, P12), which will cost more money to build. Calves produced from the beef cows (P13) can be sold (P30 and P31) or fed out and sold as finished products (P10-P12). This transfer is made possible with the use of the R23 and R24 transfer rows (Table 11.11). Labor can be hired without limitation (P21-P23) for all months except December through March. The net profitability figures, C row, reflect the nature of the particular activity, P columns, they represent. For corn production (P01) the C coefficient is negative because the corn yield can be fed to the livestock (P13) or sold through another activity (P14), and thus only production costs are shown. Whereas, for soybean production (P02) the C value is positive because the net profitability coefficient reflects gross sales

TABLE 11.9. Unit budgets and resource limitations for producing two products

Item	Corn Unit budget (1 A)	Corn Maximum production (bu)	Soybeans Unit budget (1 A)	Soybeans Maximum production (bu)
Yield (bu)	95	...	33	...
Resource requirements (limitation):				
Land (100 A)	1	9,500	1	3,300
Labor				
Apr. (80 hr)	0.3	25,365	0.9	2,937
May (80 hr)	0.8	9,500	0.5	3,762
June (80 hr)	0.7	10,857	0.4	6,600
Oct. (80 hr)	1.1	6,935	0.4	6,600
Nov. (80 hr)	1.5	5,035	...	...

Capital requirement (variable) (limitation $4,000):

	Corn Fixed	Corn Variable		Soybeans Fixed	Soybeans Variable	
Seed, fertilizer, chemicals	...	$28.50		...	$18.20	
Machinery and fuel	$ 5.75	5.36		$ 5.39	3.80	
Harvest and storage	4.80	11.70		1.65	4.83	
Labor	5.50	7.90		3.00	6.82	
Overhead	34.35	...		32.15	...	
Total/A	$52.12	$54.81	6,935	$42.19	$35.29	3,762
		$104.93			$77.48	
Total/bu	$ 0.50	$ 0.55		$ 1.28	$ 1.07	
		$1.05			$2.35	

Unit profitability:

Corn at $1.10/bu	$52.25	$ 0.55				
Soybeans at $2.40/bu				$43.89	$ 1.33	

TABLE 11.10. Alternative production levels

Alternative	Corn	Soybeans	Substitution ratio* Δ Corn	÷	Δ Soybeans	=	Ratio
A	5,066	0					
			0		1,000		0
B	5,066	1,000					
			2,566		1,400		1.83
C	2,500	2,400					
			1,000		400		2.50
D	1,500	2,800					
			1,500		137		10.95
E	0	2,937					

*The symbol Δ stands for change.

TABLE 11.11. Partial linear programming model for the complete farm budget farm

C Row		Unit	B Column Restraint	P01 Corn	P02 SB	P04 Oat-meadow meadow	P09 Swine, new	P11 Steers, old	P13 Beef cows	P18 Hay harvest	P22 Labor hired
C Row	Net profitability*	$		-47.58	66	-39.57	623	322	-15	-4	-1
R01	Land I	A	180	1	1	3					
R02	Land II	A	150								
R03	Land III	A	58								
R04	Swine facilities, old	2 litters	20								
R05	Swine facilities, new	2 litters	20				1				
R06	Beef feeding facilities, old	1 feeder	100					1			
R07	Beef feeding facilities, new	1 feeder	100								
R08	Labor, Dec.-Mar.	1 hr	1,050	0.2	0.2	0.5	11.0	3.2	4.8	0.1	
R09	Labor, Apr.-June	1 hr	1,020	2.1	2.0	0.7	5.5	2.0	2.8	2.7	
R10	Labor, July-Aug.	1 hr	600	0.4	0.9	1.1	4.4	1.6	0.9	3.8	-1
R11	Labor, Sept.-Nov.	1 hr	945	1.8	0.9	0.2	5.0	1.2	1.5	0.2	
R12	Beef cow capacity	ea	50						1		
R13	Harvested corn yield	bu	0	-100			205	62	2		
R14	Harvested oat yield	bu	0			-65					
R15	Standing meadow yield	T	0			-7.2		0.5	3.8	1	

TABLE 11.11. (continued)

	Unit	B Column Restraint	P01 Corn	P02 SB	P04 Oat-meadow meadow	P09 Swine, new	P11 Steers, old	P13 Beef cows	P18 Hay harvest	P22 Labor hired
R16 Hay transfer	T	0					0.8	1.8	−1	
R17 Hog concentrate	cwt	0				22	2.3			
R18 Beef concentrate	cwt	0						1		
R19 Annual capital costs	$	0				27		4		
R20 New investment, beef	$	0								
R21 New investment, swine	$	0				390				
R22 Set aside equality	A	50								
R23 Calf transfer, steers	ea	0					1	−0.45		
R24 Calf transfer, heifers	ea	0						−0.29		
R25 Cow transfer	ea	0						−0.16		
R26 Soybean maximum	A	0	−1	1						

*Development of net profitability illustrations:

Corn (1 A)	
Seed	$ 5.43
Fertilizer	17.60
Chemicals	2.00
Custom harvest	8.00
Shelling-drying	4.75
Machinery	8.45
Insurance	1.35
	$47.58

Swine (2 litters)	
Breeding	$ 3.50
Veterinary	18.80
Machinery	37.30
Miscellaneous	3.40
	$ 63.00
Total sales	$686.00
Net sales	$623.00

Steer feeding (head)	
Veterinary	$ 3.50
Machinery	6.50
Miscellaneous	1.00
Hauling	3.00
	$ 14.00
Total sales	$336.00
Net sales	$322.00

TABLE 11.12. List of production, purchasing, and selling activities for the complete farm budget farm

P01	Corn production, harvested and stored—Land I
P02	Soybean production, harvested and sold—Land I
P03	Oat production, harvested and stored—Land I
P04	Oats followed by 2 years meadow for pasture or hay—Land I
P05	Two years corn, 1 year soybeans, 1 year oats seed to 2 years meadow—Land II
P06	Continuous pasture—Land III
P07	Set aside—Land II
P08	Two litters swine production in old facilities, product sold
P09	Two litters swine production in new facilities, product sold
P10	Steer feeding in old facilities, product sold
P11	Heifer feeding in old facilities, product sold
P12	Steer feeding in new facilities, product sold
P13	Beef calf production, calves can be fed out or sold
P14	Selling corn—1 bu
P15	Buying corn—1 bu
P16	Allows oats to be substituted for corn in the ration
P17	Selling oats—1 bu
P18	Hay harvesting from standing meadow
P19	Buying hog feed supplements
P20	Buying cattle feed supplements
P21	Hiring labor for Apr.-June
P22	Hiring labor for July-Aug.
P23	Hiring labor for Sept.-Nov.
P24	Selling labor for Dec.-Mar.
P25	Selling labor for Sept.-Nov.
P26	Annual costs for adding new swine and beef facilities
P27	Investment capital accounting for building beef feeding facilities
P28	Investment capital accounting for building swine facilities
P29	Selling cull cows from P13
P30	Selling steer calves from P13
P31	Selling heifer calves from P13
P32	Buying steer calves to feed, P10 or P12
P33	Buying heifer calves to feed, P11

TABLE 11.13. The restraints imposed upon the complete budget farm program in the linear programming analysis and the corresponding shadow prices

Restraining situation	Row name	Limitation	Shadow prices*
1. Land I capable of producing continuous row crops	R01	180 A	$ 56.38
2. Land II must be planted to soil conserving crops	R02	150 A	44.23
3. Land III continuous pasture	R03	58 A	23.07
4. Swine produced in old facilities	R04	40 litters	197.37
5. Swine produced in new facilities	R05	40 litters	177.90
6. Beef fed in old facilities	R06	100 head	34.75
7. Beef fed in new facilities	R07	100 head	25.75
8. Operator productive labor in Dec.-Mar.	R08	1250 hr	2.00
9. Operator productive labor in Apr.-June	R09	1090 hr	2.00
10. Operator productive labor in July-Aug.	R10	620 hr	2.00
11. Operator productive labor in Sept.-Nov.	R11	1060 hr	2.00
12. Set aside acres required on Land II (item 1)	R22	50 A	8.77
13. Soybean acres limited to no more than corn acres on Land I (item 1)	R26	90 A	1.62

*Value added to net income if one more unit of the restraining resource were made available for production at its highest use.

TABLE 11.14. Activities entering the optimum farm plan as determined by linear programming with comparisons to the complete farm budget

Item	Unit	Linear programming solution	Complete budget allocation
Crop Activities			
Corn: Land I	A	90	
Land II	A	33	
		123	140
Soybeans: Land I	A	90	
Land II	A	17	
		107	65
Oats: Land II	A	17	25
Meadow: Land II	A	33	50
Set aside (held constant)	A	50	50
Pasture: Land III	A	58	58
		388	388
Livestock Activities			
Swine, farrow and finish	2 litters	40	40
Cattle feeding:			
Steers	head	200	100
Heifers	head	0	50
Beef cows	head	13	24
Steer calves fed out	head	6	10
Heifer calves sold	head	4	0
Livestock Feeds Purchased and Sold			
Haymaking	T	183	170
Corn purchased	bu	8,626	1,625
Oats sold	bu	1,083	0
Concentrate purchased:			
Hogs	cwt	880	880
Cattle	cwt	473	318
Labor Purchased and Sold			
Hire in Apr.-June	hr	255 ⎱	
Hire in July-Aug.	hr	287 ⎰	300
Sold in Dec.-Mar.	hr	18 ⎱	
Sold in Sept.-Nov.	hr	187 ⎰	500
Capital Investment Accounting			
For swine facilities	$	7,800	7,800
For beef facilities	$	660	2,300
Net Farm Income			
Value of the program	$	40,155	
Fixed costs:			
Taxes	$	2,923	
Interest	$	10,920	
Insurance	$	680	
Depreciation:			
Machinery	$	3,019	
Building and equipment	$	2,042	
		19,584	
Net farm income	$	20,571	19,266
Net farm increase	$	1,305	

minus production costs since there is not a separate selling activity established.

Note the nature of the coefficient and the development of net profitability figures illustrated at the bottom of Table 11.11. The most reliable source for this particular situation would be the records of the farm (farmer) being programmed.

The restraints placed upon the program and the corresponding shadow prices (the addition to net income that one more unit of the restricted resource would give) are shown in Table 11.13. The shadow price for Land I (R01) would tell the operator that if he could obtain one additional acre and use it at its highest and best use, $56.38 would be added to the value of the program (income over variable costs). If the operator could, or was willing to, farrow two more litters of swine per year (R05), these would add $177.90 to net income. Additional facilities for feeding out steers (R07) would return $25.75 per head at the margin.

A summary of the activities that entered the solution are shown in Table 11.14. The quantities used in the complete farm budget are shown opposite the linear programming solution. It should be realized that certain restraints were placed upon the linear program that kept it within selected limits of the budget program. For example, soybean production was limited to no more acreage than corn production and came in at this limit on Land I, which was above the budgeted acres by 42. The shadow price for soybeans was $1.62 (Table 11.13). It is interesting that all the cattle fed were steers. In fact, the solution indicated it was more profitable to sell the heifers raised and buy steers than to feed the heifers out. Steer feeding came in at the limit of 200 head, 50 head more total feeders than in the budget solution.

The linear program was developed to show the quantity of livestock concentrate to be purchased, the amount of labor to be hired, and the value of capital investment in new facilities. These are all shown in Table 11.14. Such figures are very useful to the operator in planning purchases, sales, and credit needs.

The value of the linear program, after adjusting for fixed costs not included, was $1,305 higher than the budgeted net income. This is a substantial amount but may not be as significant as some of the other management indicators shown. A study of results compared to the information going into the linear program would be a profitable exercise.

This brief discussion of management principles and techniques comes at the conclusion of a detailed treatment of farm records and accounts. The emphasis in this chapter appropriately was on the use of records in farm planning rather than on management principles and techniques. Records and accounts are tools to be used by the farm planner. The reader now should have a greater appreciation of what records and accounts to keep and how to use them for his economic benefit.

12

FARM FAMILY
LIVING ACCOUNTS

ALTHOUGH MOST FARM FAMILIES keep a fairly complete set of farm financial records because of income tax reporting requirements, relatively few families keep a complete and accurate family living expenditure record. An accurate record of this kind can be an important tool in financial planning and budgeting for the farm family. This is especially true for the young family that often has limited capital resources and reserves and may have very limited income in some years. Even under capable management farm income can vary from year to year due to factors such as the prices and weather that are beyond the control of the manager.

Most farm families could profit by keeping and analyzing their living expenditures monthly and annually. Too many consider that since detail of living expenditures is not needed for tax purposes, there is no need to keep such a record. Although most living expenditures are not tax deductible, some may be when a family itemizes deductions on schedule A (Form 1040) of the individual income tax return. Thus a family should keep an accurate record of potential tax-deductible items because they may not know in advance whether deductible expenditures will be a large enough total to make itemizing justifiable.

Some of the items that may be deductible for income tax purposes are as follows: contributions or gifts to certain religious, charitable, educational, or literary organizations; contributions to certain scientific and veterans' organizations; interest on personal notes, house mortgage, or life insurance loan; medical and drug expenses if they exceed certain specified limits; certain losses and other expenses as defined by federal and state tax laws. Families should keep informed of current tax regulations so they can capitalize on potential deductions to lower their taxable income.

Each year a family faces the decision of how to use the net income produced in the business. Expenditure patterns that are already formed and business commitments may greatly reduce the flexibility of how income might be spent in any one year. Net income can be used in several ways:

1. Pay for family living expenses. There can be a wide variation in living expenditures depending upon the values and goals of the individual family. The amount they spend

249

for living may be dependent upon their net income or it may be relatively independent. This is especially true in the short run when families may not adjust to
changes in net income. Families with relatively low income tend to spend a high
percentage of their income for living expenses because these costs are rather fixed.
When income increases, living expenses usually rise, but not as fast as income;
therefore, a smaller percentage of the income is used for living.

2. Expand the business. Many farm families depend upon capital generated within the
business to finance further expansion and to increase potential future income. The
capital expansion may come entirely from net income over and above the living costs
of the family.

3. Reduce debts or increase equity in the business. Families who are borrowing money
for business purposes may wish, as one of their goals, to increase their equity by
reducing this debt and to maintain the business assets at current levels.

4. Invest outside the business. Families may decide they wish to have some of their
earnings invested outside the business if there is adequate capital available. They
may feel this would spread their investment into several areas rather than just in the
farm business. They may have adequate capital invested in their farm business and
wish to make outside investments to diversify or because they feel that investment
opportunities are more attractive outside.

5. Increase protection or security by purchasing life insurance or a retirement annuity.
Some farm families may like to increase life insurance protection or build up a
retirement annuity from income. This may be also considered an investment outside
the business or an investment to minimize risk.

It is not unusual for farm families to find that income is not adequate to achieve all
their goals and objectives as rapidly as they would like. Therefore, they are forced to
compromise and to establish priorities. Families who know their expenditure pattern are
in a much better position to establish realistic priorities than families who do not. The
more information the family has relative to their business, financial position, and expenditures, the better position they will be in to make wise decisions relative to family
living expenditures, business investments, and financial planning.

The main purpose of keeping family living expenditure records is the asistance they
can provide when budgeting for the future. Keeping a record of family living expenditures may not help a family spend less but it may assist in developing a wiser
expenditure pattern and obtaining a higher standard of living with the same expenditures, or it may help them enjoy the same standard of living at less cost. At times
families may be forced to hold their living expenditures at certain levels because of
limited income or their current financial situation, or their lender may insist they keep
living expenditures within certain limits to protect a loan. For these families a family
living expenditure record is a must. Many farm families are not under pressure to hold
down living expenses, but an expenditure record is useful to indicate how their money is
being spent and to help them make wiser decisions relative to total expenditures.

A farm family may find it helpful to analyze how their expenditure patterns compare
with others. Only the individual family can decide how much they can and should spend
for living and how much to spend in the different categories. Table 12.1 indicates how
Iowa farm families allocated their living costs in 1974 as a percentage of their
total family living expenditures. This pattern has been much the same for the past
several years. Farm families spend a higher proportion for food than any other category.

During the past decade the percentage for food has declined. The percentage for clothing and personal items, recreation, education, and health has increased during the same period. Housing expenditures are considerably lower for farm families than for nonfarm families. Table 12.2 reports the expenditures for farm families by years during the period 1971-74.

In 1974 the farm families supplying the living expenditure data spent a total of $9,908. This is slightly over $835 a month. Many families do not recognize that their expenditures are this high unless they keep some type of record. It is also interesting to note that expenditures have increased each year even though net farm incomes have varied considerably. As reported in Table 12.2 the net income of the families in the study dropped 55 percent in 1974 but increased 77 percent in 1972 and 116 percent in 1973. At the same time the expenditures of these families increased 3 percent in 1972, 19 percent in 1973, and 13 percent in 1974. Living expenditures of individual families tend to increase year after year for several reasons:

1. Rising price levels due to inflation and other factors require a higher expenditure to maintain the same standard of living.
2. As families strive to achieve a higher standard of living, they spend more money. In some cases it may be a matter of trying to keep up with the Joneses.
3. An increase in family size normally results in increased expenditures. However, expenditures do not go up as rapidly as family size increases, so that the expenditures per person are actually less in larger families than in smaller families even though the total expenditures of larger families are greater.
4. Changes in the family life cycle will change expenditure patterns within the family. Expenditures tend to increase as the children become older. The largest increases are for clothing and personal items, food, education, recreation, and automobile. Likewise, when the children leave home or graduate from high school and college, expenditures may decline as these members become self-supporting.

One of the complaints of families concerning family living records is that they are difficult to keep since many of the purchases are paid in cash and all members of the family have expenditures. However, many of these may be relatively small but are frequent enough that they may become quite substantial in total throughout the year.

TABLE 12.1. Percentage distribution of Iowa farm family living expenditures in 1974

Expenditure category	Percentage of expenditures
Food, purchased	22.2
Food, farm raised	2.8
Clothing and personal items	13.1
Household operation and repair	9.3
Home improvements and furnishings	12.2
Health	9.6
Recreation	5.6
Education	2.9
Giving	6.9
Auto, operative and new*	7.2
Life insurance	8.2
Total	100.0

*Personal or nonbusiness share.

TABLE 12.2. Comparison by years of farm family living expenditures

Item	1974	1973	1972	1971
Cash expenditure for living:				
Food purchased	$ 2,197	$ 1,894	$ 1,674	$ 1,649
Clothing and personals	1,302	1,189	1,092	928
Household operations	574	604	463	447
Repairs	344	285	217	211
Health	953	917	787	748
Recreation	558	419	384	377
Education	284	293	236	325
Giving	683	661	460	412
Auto, operative	626	424	447	376
Total Cash Living Expense	$ 7,521	$ 6,686	$ 5,760	$ 5,473
Investments for living:				
Home improvement	$ 573	$ 265	$ 158	$ 400
Home furnishing	638	562	423	310
Auto	83	271	101	132
Total Investments for Living	$ 1,294	$ 1,098	$ 682	$ 842
Life insurance	811	654	624	620
Total Cash Expenditures	$ 9,626	$ 8,438	$ 7,066	$ 6,935
Farm produce used	282	330	282	228
Total Used for Living	$ 9,908	$ 8,768	$ 7,348	$ 7,163
Net farm income (accrual basis)	$21,467	$47,691	$22,058	$12,455
Available for income tax and investment	$11,559	$38,923	$14,710	$ 5,292
Size of family	4.07	4.27	4.57	4.5
Size of farm (acres)	446	412	396	392
Percent owners or part owners	66%	64%	57%	68%
Number of families	217	172	161	244
Percent income used for living	46%	18%	33%	58%

There are several methods families can use that may assist them in keeping complete records more easily and accurately. These techniques are as follows:

1. Pay for items by check. Paying by check provides a listing of the purchase either on the check stub or the check itself. This serves as a record of information to be entered in the proper place in the family living expenditure record book. Filling out details on check stubs such as check number, date, amount, to whom paid, and item purchased can be helpful in listing details in the family living accounts.
2. Use a credit card for purchases. If payment is made as monthly statements are received, there is no interest or service charge. Therefore, a record of purchases can be obtained with a minimum of direct cost.
3. When making cash purchases request itemized receipts. These are often given as a pattern of business, for example, from grocery, clothing, and hardware stores. Some stores such as auto service stations give receipts only upon request.
4. Make daily entries or notes of cash purchases while it is still easy to remember.
5. Place children on a cash allowance where they make many of their own purchases and keep their own records. This may include money for clothing, recreation, gifts, education, and school lunches.
6. Have a specified location (drawer, file folder, box, or nail) where receipts and expenditure notes can be held until there is time to post them. Bring postings up to date

weekly, biweekly, or (as a minimum) monthly. This activity may coincide with bill paying.

Having all family members account for expenditures they have made, either in a master record book or individual records, will provide the total family data. It will be necessary to combine all expenditures into a master book monthly or annually to get a total if individual records are maintained. Canceled checks or the credit card statement will furnish evidence that payment has been made. It is also wise to keep the purchase receipts when a warranty or guarantee is made or where a return of the item may become necessary for any reason. It is wise to establish a file for receipts, warranties, and sales slips so that they are readily available when and if needed. It is also advisable to have some type of record or evidence of contributions and other possible income tax deductions. This record may be in the form of a canceled check, receipt, or statement.

Families should give some thought as to what type of detail they need for their purposes. For instance, purchases at grocery stores normally include cleaning supplies, pet supplies, and miscellaneous items in addition to food. If families want a very detailed breakdown in their account book, they may find it necessary to separate non-food from food purchases and enter them into the proper categories. If they do not need this detail, they may enter it all as food, recognizing that a percentage of this was not specifically for food. The total expenditure will not change but the percentage that will end up being classified as food would be higher than if the other items were broken out. Another decision a family may wish to make is whether an expenditure breakdown for clothing and personal items is necessary for each member of the family or if only a family total is deemed necessary.

Many families start out keeping a living expenditure record in good faith but only complete part of the year. Since expenditure patterns are often fairly uniform from month to month, a record for several months may help to identify the expenditure pattern of the family. In this type of approach it should be recognized that many major expenditure items do not come monthly. Payments on insurance premiums, income and property taxes, social security, and other items may be made quarterly, semiannually, or annually. If several of these payments fall in the same month, this should be considered when budgeting.

Many good record books are available in which to list family living expenditures.* Most state cooperative extension services have family account books available at a nominal fee. It is also very easy for families to develop their own record books from loose-leaf or bound notebooks, using their own headings and breakdowns. Tabulation sheets of the size desired can be ruled or purchased at a bookstore. Forms 12.1 and 12.2 show some sample sheets (with entries) that families might develop for their own use.

One of the advantages in keeping a family expenditure record is the information it provides in budgeting for the future. Form 12.3 shows a sample budget a family might use for reference in estimating their expenditures for the year. The previous year's record will give an excellent guideline as to average expenditures for food and other categories. A family may use these averages to arrive at monthly budgets. They can then budget the major items that may influence the expenditure pattern the most and then can sum this to get an idea of the expected expenditure pattern for the year.

It is good management to pay for all potential tax deductible items by check to have

---

*Two excellent family account books are *Your Family Finances,* Iowa State Univ. Press, Ames, and Family Living Account Book, South Dakota Coop. Ext.Serv. Bull. 544, Brookings.

Form 12.1. Expenditures for Month of _January_

Date	Expenditure	Food	Clothing and personal	Operating expense	Repairs	Medical and health	Recreation	School and education	Church, charities, gifts	Auto
2	Groceries	$24.63								
3	Church								$8.00	
4	Electric bill			14.93						
6	Repair				$8.24					
8	Dental					$6.00				
9	Dress		$32.96							
9	Dinner	6.18								
9	BB game						$3.00			
12	School supplies School							$2.34		
14	Lunch	3.00								
15	Auto license									$36.00

Form 12.2. Example Family Living Expense Record

Food[a]

Date	Detail of purchase	Amount
1/2	Groceries, Bill's Market	$ 24.63
1/9	Dinner, Joe's Cafe	6.18
1/14	School lunch ticket	3.00
1/14	Groceries, Bill's Market	33.12
1/24	Groceries, Bill's Market	17.24
	Total	

[a]Use separate sheet for each expenditure category needed.

Form 12.3. Expenditure Budget for 19 74

Expenditure category	Total past year	Jan.	Feb.	Mar.	Apr.	May	June	July	Aug.	Sept.	Oct.	Nov.	Dec.
		Budgeting expenditures and payments											
Food	$1560	140	140	140	140	140	140	160	140	140	140	150	160
Clothing Personals	960	70	70	70	80	80	80	80	130	90	80	70	90
Operating	425	35	35	35	35	35	35	35	35	35	35	35	40
Repairs	210	20	20	20	20	20	20	20	20	20	20	20	20
Medical	690	60	60	60	60	60	60	60	60	60	60	60	60
Recreation	930	20	20	20	20	20	20	180	50	20	20	20	40
Education	285	20	20	20	20	20	5	5	5	100	20	20	20
Giving	420	35	35	35	35	35	35	40	40	40	40	40	40
Auto non bus	385	50	25	25	60	25	25	25	25	25	60	30	30
Home imp	385	10	10	200	10	10	10	10	10	150	10	10	10
Home furn	380	15	20	20	20	20	20	20	20	20	250	20	20
Life ins	550		45		270			165			70		
Income taxes	320		430										
Total	6900	475	930	645	770	465	450	800	535	700	805	475	530

Total estimated for the year $7580

a record of the payment. Canceled checks also provide evidence that a bill or account has been paid. A family may also like to star possible tax-deductible items in a record book if they itemize deductions. It is then possible to thumb through the record book and get a quick summary of the deductible items such as contributions, medical expenses, etc.

Keeping family living expenditure records and analyzing them is part of the financial planning of the farm couple. They also need an accurate record of farm expenses and income and the annual financial statement. The financial statement will provide information on the net worth change the family experienced during the year. If the family living expenses are as large as the net income, they will have no net worth increase.

13

THE FARM BUSINESS CENTER

FARMING IS A BUSINESS, often very large in terms of capital managed and amounts of money handled. Modern farm businesses spend sizable amounts of money for items such as machinery, equipment, feeder cattle, feed, fertilizer, seed, chemicals, and other supplies. Farmers also spend for services such as trucking, labor, machine hire, and repair services. They also handle a considerable amount of capital as income from the sale of crops, livestock, and livestock products.

In operating a farm, there are many business associated activities that must be handled if the farm business is to be managed successfully. These include maintaining a complete and accurate accounting of receipts and expenditures for the following purposes:

1. Provide the information necessary to file accurate income tax reports.
2. Furnish data to prepare financial statements and cash-flow budgets.
3. Provide the basic data for analysis of the farm business.
4. Provide budgeting information for use in planning.
5. Maintain a record of accounts payable and accounts receivable.

Certain activities associated with a successful farm business can best be handled in a working space such as a farm business office provided specifically for this purpose. These activities include:

1. Listing receipts and expenditures of the farm business.
2. Keeping current on the market situation by use of radio, TV, daily newspapers, and newsletters.
3. Conducting business activities by phone, such as buying and selling, obtaining services, and checking or obtaining market quotations.
4. Collecting and computing information needed in the farm business.
5. Keeping up to date or current on modern technology that may be applicable to the business.

6. Studying operating manuals to obtain the knowledge needed for the most effective machine operation.
7. Maintaining correspondence and writing checks to pay bills.
8. Filing and storing information needed for future reference.

To carry on the above activities associated with the farm business, adequate space in comfortable surroundings is desirable. The amount of space that can or should be allocated to a farm business office will vary with the alternatives available to the individual family. The ideal situation would be to have an office space of adequate size as part of the family dwelling. Preferably this will be located so that access is available directly from the outside or from a service entryway. Location should be near the entrance that the operator uses most in his daily farming routine. There may also be an inside office entrance from the rest of the family home. Preferably the location should be such that many farm activities can be observed from the office. Provision should be made for adequate heat and ventilation so that it can be used daily throughout the year as needed.

Adequate and efficient communication is necessary for successfully running a farm business. Communication facilities that may be considered for the office are as follows:

1. An extension phone in the office or a business phone with a listing separate from the family.
2. A two-way intercom system from the office to key farm activity areas such as the farm shop, feed handling center, or livestock care centers such as the farrowing house or milking parlor.
3. A short-wave radio or short-range radio to the pickup truck, tractors, combines, and other key machinery or transportation vehicles to maintain communication with the units when they are engaged in field or transportation activities.
4. An intercom system with other parts of the house to facilitate communication with other members of the family.

Many families may not have separate space available for a farm business office; therefore, they may arrange to provide space in the kitchen, family room, utility room, living room, or in a basement room or closed-in porch. Much will depend on the needs of the individual business and the space alternatives available. The costs associated with building, equipping, and maintaining the farm business office can be treated partially or totally as a business expense and should be handled thus for tax purposes. If it is not used completely for farm business activities, the costs would be prorated to reflect the amount of business use. The costs would be handled in a tax report much as those for other machinery, equipment, and buildings.

The space available in the farm business office will determine the amount of equipment and furniture that can be used. If the space is limited, equipment and furniture may also be at a minimum. Basic equipment to maintain a satisfactory farm business office should include a desk, a desk chair, one four-drawer file or two two-drawer files, a wastebasket, and a desk lamp. Additional equipment might include an adding machine, a typewriter, bookshelves and storage area for farm magazines, a radio, a television set, a fireproof safe, and a bulletin board. The personal desires of the manager and the family will bring modifications or additions to this list.

The farm business office also requires certain office supplies for handling incoming and outgoing mail, financial records, educational materials, filing, etc. A suggested list of supplies includes:

1. Standard manila file folders, and assorted other file folders
2. Stapler
3. Appointment calendar
4. Paper clips, thumbtacks, rubber bands, and other desk supplies
5. Stationery, postcards, and assorted stamps
6. Receipt books
7. Pencils, pens, erasers, etc.
8. Farm and home account book
9. Scratch pads, clipboard, etc.

A visit to an office supply store will suggest articles that are available and may be helpful.

The farm business office may also be used as an information center in which to find suggestions for the most profitable operation of the farm. Many times the operator may want some specific instructions on insect control, disease control, proper use of chemicals, or balanced rations for livestock. A partial list of the educational and informational materials a farm operator may find useful in his office follows:

1. Machinery and equipment operating manuals
2. Farm and business magazines or periodicals
3. Papers providing market information
4. Textbooks on selected topics
5. Bulletins from agricultural experiment stations, agricultural extension services, or business sources
6. Situation reports such as USDA reports, market news reports, and newsletters

The key to an efficient office is a well-organized and adequate filing system that assists in locating information quickly and with a minimum of effort. A suggested filing guide is shown in Form 13.1. Place a filing guide number on each item so that it is not misfiled originally or when removed at a later date. Once this is done, any member of the family can assist with filing without difficulty. The number also makes it easier to spot an item that may have been misfiled or misplaced.

Prepare several copies of the filing system guide that you plan to use. Keep one copy at the desk for reference and another copy in the file. The filing system will be more difficult to use at first than a less organized system. However, in time many of the filing system numbers will be memorized. The filing system can be easily adjusted to fit individual situations.

The filing system illustrated in Form 13.1 is established on a name and numbers system that should make items easy to locate when needed. The system is divided into seven major sections:

Section I.   Farm business
Section II.   Home and family living

Form 13.1.   Codes for Farm Business Filing System

Suggested Farm Business Filing System

Section I Farm Business

1.0 Business record account book
2.0 Expenditure slips (Enter in record book and then file in envelopes by months; combine at end of year)
3.0 Open charge accounts (Enter in record book then transfer slips to 2.0 when accounts are paid)
4.0 Gasoline purchases (Enter in record book and keep for gasoline tax refund purposes)
5.0 Hired labor records (Enter in record book and keep tally sheet and social security information for each employee)
6.0 Livestock purchases (Enter in record book before filing)
7.0 Sales (Enter in record book before filing)
    7.1 Cattle sales
    7.2 Hog sales
    7.3 Dairy product sales
    7.4 Egg and poultry sales
    7.5 Other livestock sales
    7.6 Crop sales
    7.7 Refunds and dividends
    7.8 Government payments
    7.9 Other sales and receipts
8.0 Agreements or orders for future sales or purchases
9.0 ASCS commodity loan records
10.0 ASCS approvals and correspondence
11.0 Tax records and receipts
    11.1 County
    11.2 Federal and state income
    11.3 Social Security tax records and forms
12.0 Insurance records°
    12.1 Fire
    12.2 Liability
    12.3 Auto
    12.4 Life
    12.5 Other
13.0 Rental agreements and lease forms
14.0 Notes and loans°
15.0 Deeds and mortgages°
16.0 Awards and recognitions
17.0 Business correspondence
18.0 Bank deposit slips
19.0 Cancelled checks
20.0 Other

Section II Home and Family Living**

25.0 Home account book
26.0 Expenditure receipts
27.0 Inventory of furniture and equipment

28.0 Valuable papers for family, school records, birth records, health records, etc.
29.0 Addresses
30.0 Correspondence
31.0 Farm and family organization records
32.0 Automobile records, title, guarantees, etc.
33.0 House and garden bulletin file
34.0 Household equipment — operating instructions and manuals (file in one folder alphabetically)
35.0 Other

Section III Management and Outlook

40.0 Budgets
41.0 Plans
42.0 Market letters
43.0 Outlook reports
44.0 Other

Section IV Crop Production

50.0 Conservation plans and soil map
51.0 Soil test reports
52.0 Field records
        (rotation—yields—treatment)
53.0 Seed test records and certification tags
54.0 Weed, insect and disease control material
55.0 Crop bulletin file
    55.1 Corn production
    55.2 Soybean production
    55.3 Crop storage

Section V Livestock Production

65.0 Registration papers
66.0 Breeding records
67.0 Livestock feeding records
68.0 Livestock bulletin file (file by livestock classes)

Section VI Machinery and Buildings

75.0 Operator's manuals (file in alphabetical order)
76.0 Conditional sales contracts
77.0 Farm building plans
78.0 Farmstead layout plans
79.0 Farm equipment plans
80.0 Machinery and building bulletin file

Section VII Inactive Business Records

85.0 Records books for past years
86.0 Business receipts (at least past three years)
    86.1 Sales and income
    86.2 Purchases and expenditures
87.0 Cancelled checks (at least past three years)
88.0 Deposit slips (at least past three years)
89.0 Income tax reports (at least past three years)
    89.1 Federal tax reports
    89.2 State tax reports

---

* These should be kept in bank safety deposit box or fireproof safe.

** Two Iowa State University publications can be of further assistance in this area. One of these is "Your Family Business Affairs," HE 75. It contains additional information in the following areas: (1) What valuable papers to keep, where to keep them and how long, (2) further comments about your business center, (3) and suggestions on long-range planning in the areas of wills, ownership arrangements and insurance. The accompanying publication, HE 75-W, contains worksheets for Your Family Business Affairs. Copies of these publications can be obtained by writing and requesting them from Publications Distribution Center, Printing and Publications Building, Iowa State University, Ames, Iowa 50010, or your local extension service office may have copies.

Suggested headings under these sections are shown in Form 13.1. This filing system is flexible and may be expanded by adding decimals and additional related files in any section. For example, item 6.0, livestock purchases, could be divided thus—6.1 cattle, 6.2 hogs, 6.3 sheep, etc. Numbers have been omitted between sections to permit addition of new headings in related areas.

A well-planned business office will make it easier to be a good manager. It provides the environment and facilities for record keeping, studying informational material to keep up with modern farming technology, and planning. A well-organized filing system permits orderly location and quick retrieval of data and information. Farming is a business. A farm business office is one of the tools the manager needs to operate his farm business successfully.

Several pieces of equipment can increase efficiency in handling farm business affairs. Electric adding machines cost less today than nearly any time in the past. The operator may even wish to consider getting an electric calculator. An adding machine can be used for addition, subtraction, and multiplication. Many farmers consider the purchase of an electric adding machine one of the best investments they have made. Most persons would prefer a machine that prints out the entries and the results for checking and reference. Here again individual preferences will determine the type of machine that might be most useful. A typewriter is an excellent piece of equipment to facilitate communication or correspondence if one or more members of the family have some typing skill. Consideration may be given to an electric typewriter if considerable use is made of it. It may be possible to purchase a used typewriter that meets the requirements of the individual farm business.

Farmers are faced with the decision of which business records they should keep and how long to keep them. Keeping detailed records of business transactions over time means an accumulation of considerable data. If the information is not important and will not be used in the future the question becomes, Why keep it? One of the main reasons for keeping data is for filing an accurate and complete income tax report. Since the taxpayer is always subject to audit, he must be sure he has adequate records if he is audited by the Internal Revenue Service (IRS). The audit can go back three years. In case there is evidence of fraud, the three-year limit may not apply, so that records could be checked over a longer period of time.

There may also be occasions when you realize you have made errors in your income tax report, and it may be to your advantage to file an amended return. Here again, you can go back three years. If the IRS audits you and goes back several years, you can go back three years beyond that. Therefore, keeping records over a period of time offers protection. In view of the possibility of IRS audit, it is advisable to keep farm business records for a period of at least three years; it may even be safer to keep them for five or six.

Records for all information in the account book should be kept in addition to canceled checks, deposit slips, bank statements, receipts, bills of sale, and other evidence of business transactions. It is probably advisable to keep a record of depreciable items three years beyond the time they are sold or taken out of service and removed from the depreciation schedule because of recapture on tax returns. Filing tax return information on the purchase price and depreciation taken on each machine is necessary when it is sold. For instance, if a farmer purchases a tractor today and uses it for 10 years before selling it, he must report the sale price at that time. He may have some depreciation recapture if it was depreciated below the sale price. Since the individual again is subject to audit three years beyond this point he may wish to keep the data for this tractor for 13 years from time of purchase.

The same is true of soil and water conservation expenses and land clearing expenses. If these items are deducted as expense, keep the record a minimum of ten years because of the recapture rule if the farm is sold within ten years of the time these expenses were incurred. Here again, you may wish to keep them a minimum of three years after the farm is sold. If you decide to capitalize these items rather than write them off as a business expense, the information must be kept and the cost added to the basis of the farm. Under these conditions a record of these investments should be kept as long as you own a farm.

In the case of buildings, improvements, and equipment you should keep the information as long as these items are in possession, plus three years. A current depreciation schedule is necessary during this period, but it is also advisable to keep evidence of the cost and when the item was purchased and first used. In the case of the farm dwelling which cannot be depreciated, you would keep information as long as it was owned. If major improvements are made such as air conditioning or enlarging the dwelling, this information should be kept because it is added to the basis. This information will be needed for the tax return when this item is sold.

It is also good business to keep available information on machinery and equipment. This would include instruction manuals and parts lists. It is advisable to keep these data as long as the machine or equipment is in use by the buyer. Keeping machinery information may save considerable time if it is necessary to order repairs for the machine. The instruction book may be very useful when repairs or adjustments are needed.

For farm business planning purposes it is often advisable to keep certain data as a record of past performance. Such information may include a crop record that includes fertilizer and lime applications for individual fields and the soil test for these fields. All this can be helpful when planning crop production in the future. A map indicating the location of title lines is also a good item to keep over time.

If deductions are itemized on the tax report, keeping this information for a period of three years is feasible. If the items are not deducted on the tax report, keep the information only if it is significant in the business. Information on purchase of equipment and furniture may be helpful in case there is a question about the item. Many items have a guarantee of one to five years, so that information on purchase date and cost might be helpful if this guarantee needs to be exercised.

The farmer then faces the dilemma of where to store data that are seldom used but still should be kept for the reasons given above. It may not be feasible to keep this information in files because of the bulk involved. If it is placed in storage boxes or storage facilities they should be labeled so that recovery can be as simple as possible. Also, storage boxes should be kept dry and stored in such a manner as not to create a fire hazard.

APPENDIX

TABLE A.1. Weights and measures

LIQUID MEASURE

		Metric
1 tablespoon	= 3 teaspoons	
1 ounce	= 2 tablespoons	= .0296 liter
1 cup	= 8 ounces	= .2366 liter
1 pint	= 2 cups	= .4732 liter
1 quart	= 2 pints	= .9463 liter
1 gallon	= 4 quarts	= 3.7843 liters
1 cu ft	= 7.48 gals water	=24.3066 liters

DRY MEASURE

1 quart	= 2 pints	= 1.1012 liters
1 peck	= 4 quarts	= 8.8096 liters
1 bushel	= 4 pecks	=35.2383 liters
	= 1.244 cu ft	

LINEAR MEASURE

1 link	= 7.92 inches	
1 foot	= 12 inches	= .3048 meter
1 yard	= 3 feet	= .9144 meter
1 rod	= 16.5 feet	= 5.029 meters
	= 25 links	
1 chain	= 4 rods (66 ft)	
1 mile	= 5,280 feet	= 1.6093 km
	= 80 chains	

SQUARE MEASURE

1 sq inch		
1 sq foot	= 144 sq in.	= 6.452 sq cm
1 sq yard	= 9 sq ft	= 929 sq cm
1 sq rod	= 30.25 sq yds	= .8316 sq meter
1 acre	= 160 sq rods	= 25.29 sq meters
	= 43,560 sq ft	= 4046 meters
1 section	= 640 acres	= .4047 hectare
		= 2.59 sq km
1 township	= 36 sections	= 259 hectares

CUBIC MEASURE

1 cu inch		
1 cu foot	= 1,728 cu in.	= 16.387 cu cm
1 cu yard	= 27 cu ft	= .0283 cu meter
1 gallon	= 231 cu in.	= .7646 cu meter
1 cord	= 128 cu ft	
(wood)	(8'x4'x4')	

LAND MEASURE-RECTANGULAR SYSTEM

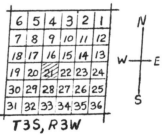

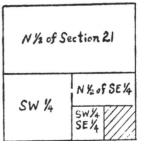

T3S, R3W

The SE¼ of SE¼ of Section 21,
T3S, R3W of the 5th P.M.

TABLE A.2.   Rules for estimating grains and roughages

1. **TO FIND THE NUMBER OF BUSHELS OF GRAIN OR SHELLED CORN IN A BIN:** Measure length, width, and average depth of the grain in the bin. Obtain total cubic feet of grain by multiplying the length by the width by the depth (all in feet). Divide by 1¼ (or multiply by 0.8) to find bushels.

2. **TO FIND THE NUMBER OF BUSHELS OF EAR CORN IN CRIB:** Rectangular crib — multiply the length by the width by the average depth (all in feet) and divide by 2½ (or multiply by 0.4) to find bushels. Round crib — multiply the distance around the crib by the diameter by the depth of the corn (all in feet) and divide by 10.

3. **TO FIND THE NUMBER OF TONS OF HAY IN MOW:** Multiply the length by the width by the height (all in feet) and divide by 400 to 525, depending on the kind of hay and how long it has been in the mow.

4. **TO FIND THE NUMBER OF TONS OF HAY IN STACK:** Rectangular stacks — secure the overthrow, O (the distance from the ground, close to stack, on one side over the top of the stack to the ground on the other side); the width, W; and the length, L (all in feet). Additional rules for round stacks will be found on page 4 of U.S.D.A. Leaflet No. 72, published in February, 1931.

   The contents in cubic feet may then be determined as follows:
   (a) For low round-topped stacks[(.52 × O) — (.44 × W)] × W × L
   (b) For high round-topped stacks[(.52 × O) — (.46 × W)] × W × L
   Divide the number of cubic feet thus secured by the following cubic feet allowed per ton:

When settled:	1 to 3 months	Over 3 months
Alfalfa Hay	485	470
Timothy	640	625
Wild Hay	600	450

   For clover hay use about the same as alfalfa or slightly higher (500 to 512). Most Iowa hay stacks come under Class (a) above.

5. **TO FIND THE NUMBER OF TONS OF STRAW:** Follow the same method as is used with hay except that about twice as many cubic feet (900 to 1000) are allowed per ton.

6. **FODDER AND STOVER** are usually estimated on the acre basis, estimating the amount of corn in the fodder and allowing some additional value per acre for the stover.

7. **TO FIND TONS OF SILAGE IN SILO:** The weights of silage in the table to the right show that has been settled at least one month. The total amount of silage in a silo can easily be determined if one knows the diameter of the silo and the depth of the silage in it.

   A. The amount of silage in a silo may be taken directly from table if no silage has been removed.

   B. To estimate amount of silage in silo one month or more after filling:
   (1) Determine original height of settled silage and read tonnage from table.
   (2) Determine depth of settled silage removed and read tonnage from table.
   (3) The tonnage remaining is obtained by subtracting (2) from (1). Example: A 14′ silo was filled to a depth of 35′ (settled height) and 16′ have been removed. Find tonnage remaining.

Total capacity before removal of silage	106.6	tons
Weight of the 16′ of silage removed	45.2	″
Amount remaining	61.4	″

   C. Suggested corrections:
   (1) If corn is put in silo while in the milk stage add 10-15 per cent to weights given in table. If grain is unusually heavy in production to stalk add 5-10 per cent. If very little or no grain is present deduct 10 per cent.
   (2) If corn is past the usual stage of maturity and contains less water than usual, deduct 10-15 per cent.
   (3) Fineness and tramping have no effect on the volume per ton of silage.

8. **SILAGE IN TRENCH SILO:** (I.S.U. Circular 202). Multiply the depth by the length by the average width (all in feet) and secure total cubic feet. Divide the total cubic feet by 60 to secure rough estimate of tons capacity, since a cubic foot of such silage weighs about 35 pounds.

NOTE: For silos filled with sorghum or hay ensilage, the above tonnage capacities should be increased 35 to 50 per cent.

LEGAL WEIGHTS PER BUSHEL

Alfalfa	60	Corn (dry ear)	70	Potatoes	60
Apples	48	Corn (shelled)	56	Red Top	14
Barley (common)	48	Corn (sweet)	50	Rye	56
Barley (hull-less)	60	Flax	56	Soybeans	60
Beans	60	Millet	50	Timothy	45
Bluegrass, Orch., Br.	14	Oats	32	Wheat	60
Clovers	60	Peas	60	Milk (per gal.)	8½

ESTIMATED WEIGHTS OF SETTLED CORN SILAGE
ONE MONTH OR MORE AFTER FILLING †
See Rule 7 for Method and Corrections

Depth of Silage in Feet	Diameter of Silo in Feet					
	10′	12′	14′	16′	18′	20′
	Tons	Tons	Tons	Tons	Tons	Tons
1	1.3	1.8	2.5	3.2	4.1	5.0
2	2.5	3.7	5.0	6.1	8.2	10.2
3	3.9	5.5	8.0	9.9	12.5	15.4
4	5.2	7.5	10.2	13.3	16.8	20.8
5	6.6	9.5	12.9	16.8	21.2	26.2
6	7.9	11.4	15.6	20.3	25.7	31.8
7	9.4	13.5	18.4	24.0	30.3	37.5
8	10.8	15.6	21.2	27.7	35.0	43.2
9	12.3	17.7	24.0	31.4	39.7	49.0
10	13.7	19.8	27.0	35.2	44.5	55.0
11	15.3	22.0	29.9	39.0	49.3	61.8
12	16.8	24.2	32.5	42.9	54.3	67.1
13	18.3	26.4	35.9	46.9	59.3	73.3
14	19.9	28.7	39.0	50.9	64.4	79.6
15	21.4	30.9	42.0	54.9	69.3	85.7
16	23.1	33.2	45.2	59.0	74.6	92.2
17	24.6	35.5	48.3	63.4	79.7	98.5
18	26.2	37.8	51.4	67.1	84.8	104.8
19	27.8	40.1	54.6	71.2	90.0	111.3
20	29.5	42.4	57.8	75.4	95.3	117.8
21	31.0	44.7	61.0	79.4	100.3	124.0
22	32.7	47.0	64.0	83.6	105.6	131.0
23	34.3	49.4	67.3	87.8	110.5	137.2
24	35.9	51.7	70.4	91.9	116.1	143.6
25	37.6	54.2	73.7	96.2	121.6	150.3
26	39.2	56.5	76.9	100.3	126.8	156.8
27	40.9	59.0	80.2	104.7	132.4	163.6
28	42.6	61.3	83.4	108.9	137.6	170.1
29	44.3	63.8	86.9	113.4	143.3	177.1
30	45.9	66.1	90.1	117.6	148.6	183.7
31	47.6	68.5	93.4	121.9	154.1	189.9
32	49.3	70.9	96.7	126.2	159.5	196.2
33	51.0	73.4	100.0	130.5	165.0	202.4
34	52.7	75.8	103.3	134.8	170.5	208.7
35	54.4	78.3	106.6	139.1	175.9	214.9
36		80.7	110.0	143.5	181.4	221.0
37		83.1	113.3	147.8	186.9	227.4
38		85.5	116.6	152.1	192.4	233.7
39		88.0	119.9	156.4	197.8	239.9
40		90.4	123.2	160.7	203.3	246.2
41			126.5	165.0	208.8	252.4
42			129.8	169.3	214.2	258.7
43			131.1	173.6	219.7	264.9
44			136.4	177.9	225.2	271.2
45			139.7	182.2	230.6	277.4
46			144.0	186.5	236.1	282.4
47			147.3	190.8	241.5	288.7
48			150.6	195.1	247.0	295.0
49			153.9	199.4	252.4	301.2
50			156.2	203.7	257.9	307.5

† This table based on data secured by Missouri and Kansas Agricultural Universities.

NOTE: Nebraska Experiment Station data indicates silo capacities as about 1/7 larger than this table (Kansas and Missouri) or practically the same as this table if measured 24 hours after filling instead of one month later.

Source: H. B. Howell, <u>Better Farm Accounting</u> 3rd ed., Iowa State Univ. Press, Ames, 1972

TABLE A.3. Compound interest tables for varying interest rates

Years	3	4	5	6	7	8	9	10
			Rate of interest	(percent per year) (dollars)				

Value of $1 (compounded annually) at the end of an interest-bearing period of n years [a]

Years	3	4	5	6	7	8	9	10
1	1.030	1.040	1.050	1.060	1.070	1.080	1.090	1.100
2	1.061	1.082	1.103	1.124	1.145	1.166	1.188	1.210
3	1.093	1.125	1.158	1.191	1.225	1.260	1.295	1.331
4	1.126	1.170	1.216	1.262	1.311	1.361	1.412	1.464
5	1.159	1.218	1.276	1.338	1.403	1.469	1.539	1.611
6	1.194	1.265	1.340	1.419	1.501	1.587	1.677	1.772
7	1.230	1.316	1.407	1.504	1.606	1.714	1.828	1.949
8	1.267	1.369	1.478	1.594	1.718	1.851	1.993	2.144
9	1.305	1.423	1.551	1.689	1.839	1.999	2.172	2.358
10	1.344	1.480	1.629	1.791	1.967	2.159	2.367	2.594
15	1.558	1.801	2.079	2.397	2.759	3.172	3.642	4.177
20	1.806	2.191	2.653	3.207	3.870	4.661	5.604	6.728
25	2.094	2.666	3.386	4.292	5.428	6.849	8.623	10.830

Value, at the beginning of the period, of $1 received after an interest-bearing period of n years [b]

Years	3	4	5	6	7	8	9	10
1	.971	.962	.952	.943	.935	.926	.917	.909
2	.943	.925	.907	.890	.873	.857	.842	.826
3	.915	.889	.864	.840	.816	.794	.772	.751
4	.888	.855	.823	.792	.763	.735	.708	.683
5	.862	.822	.784	.747	.713	.681	.650	.621
6	.838	.790	.746	.705	.666	.630	.596	.564
7	.813	.760	.711	.665	.623	.584	.547	.513
8	.789	.731	.677	.627	.582	.540	.502	.466
9	.766	.703	.645	.592	.544	.500	.460	.424
10	.744	.676	.614	.558	.508	.463	.422	.386
15	.642	.555	.481	.417	.362	.315	.274	.239
20	.554	.456	.377	.312	.258	.214	.178	.149
25	.478	.375	.295	.233	.184	.146	.116	.092

Value of a (finite) series of n annual payments of $1 (compounded annually) at the date of the last payment [c]

Years	3	4	5	6	7	8	9	10
1	1.000	1.000	1.000	1.000	1.000	1.000	1.000	1.000
2	2.030	2.040	2.050	2.060	2.070	2.080	2.090	2.100
3	3.091	3.122	3.152	3.184	3.215	3.246	3.278	3.310
4	4.184	4.246	4.310	4.375	4.440	4.506	4.573	4.641
5	5.309	5.416	5.526	5.637	5.751	5.867	5.985	6.105
6	6.468	6.633	6.802	6.975	7.153	7.336	7.523	7.716
7	7.662	7.898	8.142	8.394	8.654	8.923	9.200	9.487
8	8.892	9.214	9.549	9.897	10.260	10.640	11.030	11.440
9	10.160	10.580	11.030	11.490	11.980	12.490	13.020	13.580
10	11.460	12.010	12.580	13.180	13.820	14.490	15.190	15.940
15	18.600	20.020	21.580	23.280	25.130	27.150	29.360	31.770
20	26.870	29.780	33.070	36.790	41.000	45.760	51.160	57.270
25	36.460	41.650	47.730	54.860	63.250	73.110	84.700	98.350

Interest formulas:

[a] $V_n = V_0 (1 + i)^n$; V_0 = Value of a sum at the beginning of the interest-bearing period.

[b] $V_0 = V_n \left[\dfrac{1}{(1 + i)^n} \right]$ V_n = Value of a sum at the end of the interest-bearing period.

[c] $V_n = \dfrac{r (1 + i)^{n-1}}{i}$ i = The rate of interest.

n = The number of years in the interest-bearing period.

r = The value of one of a series of equal payments to be made annually.

TABLE A.4.   Principal and interest paid per $1 borrowed, by term of loan and by interest rate

Frequency of payments	Number of payments	Annual interest rate								
		4%	4 1/2%	5%	5 1/2%	6%	6 1/2%	7%	7 1/2%	8%
Annually	3	$1.081046	$1.091320	$1.101626	$1.111962	$1.122329	$1.132727	$1.143155	$1.153614	$1.164102
	4	1.101960	1.114975	1.128047	1.141178	1.154366	1.167611	1.180912	1.194272	1.207684
	5	1.123135	1.138958	1.154874	1.170882	1.186982	1.203173	1.219453	1.235825	1.252285
	6	1.144571	1.163270	1.182105	1.201074	1.220176	1.239410	1.258775	1.278270	1.297896
	7	1.166267	1.187910	1.209739	1.231751	1.253945	1.276320	1.298873	1.321607	1.344511
	8	1.188223	1.212877	1.237774	1.262912	1.288288	1.313898	1.339742	1.365824	1.392120
	9	1.210437	1.238170	1.266211	1.294555	1.323200	1.352142	1.381378	1.410912	1.440720
	10	1.232909	1.263788	1.295046	1.326678	1.358680	1.391047	1.423775	1.456860	1.490300
	12	1.278626	1.315994	1.353905	1.392351	1.431324	1.470818	1.510824	1.551336	1.592352
	15	1.349116	1.396707	1.445134	1.494384	1.544441	1.595292	1.646919	1.699320	1.752450
	20	1.471635	1.537523	1.604852	1.673587	1.743691	1.815128	1.887859	1.961860	2.037060
	25	1.600299	1.685976	1.773812	1.863734	1.955668	2.049537	2.145263	2.242775	2.341975
	30	1.734903	1.841746	1.951543	2.064162	2.179467	2.297323	2.417592	2.540160	2.664840
	35	1.875206	2.004466	2.137510	2.274123	2.414085	2.557179	2.703189	2.851905	3.003140
	40	2.020940	2.173726	2.331126	2.492814	2.658462	2.827749	3.000366	3.176040	3.354440
Semi-annually	6	1.071155	1.080210	1.089300	1.098425	1.107585	1.116780	1.126009	1.135278	1.144572
	8	1.092078	1.103877	1.115739	1.127664	1.139651	1.151701	1.163813	1.175992	1.188224
	10	1.113265	1.127877	1.142588	1.157397	1.172305	1.187311	1.202414	1.217620	1.232910
	12	1.134715	1.152209	1.169846	1.187625	1.205545	1.223606	1.241807	1.260156	1.278636
	14	1.156428	1.176872	1.197511	1.218344	1.239369	1.260585	1.281990	1.303596	1.325366
	16	1.178402	1.201866	1.225584	1.249554	1.273774	1.298242	1.322957	1.347920	1.373120
	18	1.200638	1.227190	1.254061	1.281251	1.308757	1.336575	1.364703	1.393146	1.421892
	20	1.223134	1.252841	1.282943	1.313435	1.344314	1.375578	1.407222	1.439260	1.471640
	24	1.268906	1.305126	1.341908	1.379247	1.417138	1.455574	1.494548	1.534056	1.574088
	30	1.339498	1.385980	1.433329	1.481533	1.530578	1.580452	1.631140	1.682640	1.734930
	40	1.462230	1.527095	1.593449	1.661260	1.730495	1.801118	1.873091	1.946400	2.020960
	50	1.591160	1.675918	1.762903	1.852046	1.943275	2.036514	2.131686	2.228750	2.327550
	60	1.726078	1.832120	1.941204	2.053201	2.167978	2.298243	2.405317	2.527620	2.652120
	70	1.866736	1.995321	2.127798	2.263953	2.403564	2.546409	2.692267	2.840950	2.992220
	80	2.012857	2.165101	2.322084	2.483474	2.648940	2.818152	2.990791	3.166560	3.345200

Source: Farmer's Handbook in Financial Calculations and Physical Measurements, Agriculture Handbook 230, USDA, 1964.

Example: After a down payment on farmland, your balance is $10,000.
It is to be amortized over 10 years by regular annual payments.
The annual interest rate is 6 percent.

a) How much will you repay lender?
b) What will be your annual payment?
c) How much interest will you pay?

Solution:
a) 10,000 x 1.358680 = $13,586.80.   The 1.358680 is found in the 6-percent column, opposite 10 annual payments.
b) 13,586.80 ÷ 10 = 1,358.68 per year.
c) 13,586.80 - 10,000 = $3,586.80 interest.

INDEX